FASTtrack

Physical Pharmacy

David Attwood
Professor, School of Pharmacy and
Pharmaceutical Sciences,
University of Manchester, UK

Alexander T. Florence
Professor Emeritus, The School of Pharmacy,
University of London, UK

Pharmaceutical Press
London • Chicago

Published by the Pharmaceutical Press
An imprint of RPS Publishing
1 Lambeth High Street, London SE1 7JN, UK
100 South Atkinson Road, Suite 200, Grayslake, IL 60030-7820, USA

(**PP**)is a trade mark of RPS Publishing

RPS Publishing is the publishing organisation of the Royal Pharmaceutical
Society of Great Britain

First published 2008

Design and layout by Designers Collective, London
Printed in Great Britain by TJ International, Padstow, Cornwall

ISBN 978 0 85369 725 1

A catalogue record for this book is available from the British Library.

FASTtrack

Physical
Pharmacy

Contents

Introduction to the *FASTtrack* series

FASTtrack is a new series of revision guides created for undergraduate pharmacy students. The books are intended to be used in conjunction with textbooks and reference books as an aid to revision to help guide students through their exams. They provide essential information required in each particular subject area. The books will also be useful for pre-registration trainees preparing for the Royal Pharmaceutical Society of Great Britain's (RPSGB's) registration examination, and to practising pharmacists as a quick reference text.

The content of each title focuses on what pharmacy students really need to know in order to pass exams. Features include*:

- concise bulleted information
- key points
- tips for the student
- multiple choice questions (MCQs) and worked examples
- case studies
- simple diagrams.

The titles in the *FASTtrack* series reflect the full spectrum of modules for the undergraduate pharmacy degree.

Titles include:
Pharmaceutical Compounding and Dispensing
Managing Symptoms in the Pharmacy
Pharmaceutics: Dosage Form and Design
Pharmaceutics: Delivery and Targeting
Therapeutics
Complementary and Alternative Therapies

There is also an accompanying website which includes extra MCQs, further title information and sample content: www.fasttrackpharmacy.com.

If you have any feedback regarding this series, please contact us at feedback@fasttrackpharmacy.com.

*Note: not all features are in every title in the series.

Preface

University education is about acquiring knowledge, not just passing examinations. The latter is of course a necessary marker of progress, and as a result many students feel that examinations dominate their academic years. In our experience, as teachers of pharmaceutics and physical pharmacy over many years, many students do not adopt proper modes of study for any of their subjects, or even devote enough time to revision. No textbook substitutes for students' own notes gathered during lectures, annotated later with more information from textbooks and a timetable of revision leading up to examinations. Reading notes at intervals of days is much more effective than re-reading the notes several times in one day.

The *FASTtrack* series is intended not as an alternative to textbooks but as an aid to revision, providing the key points of each topic and questions with which progress in learning can be gauged. But, like past examination papers, these can only give clues as to what might come in the examination which you are to sit. What you must always ask as a user of this book is: what kind of question might I be asked about topic A and topic B?

This book is derived unashamedly from the fourth edition of our textbook *Physicochemical Principles of Pharmacy*, published by the Pharmaceutical Press in 2006. It is not a substitute for it but should be used alongside it for those revision periods when time is short. In many cases you will need to refer to the full text for more detail. You will find fewer equations here. How can we have a physical pharmacy text without equations? There are few drug structures, yet an understanding of structures is essential for understanding physical pharmacy, formulation and drug behaviour. Hence for a complete understanding of some areas you must refer to drug structures. In examination answers it is important to include appropriate drug structures, equations and diagrams not only but especially in this subject. *Physicochemical Principles of Pharmacy* has structures, diagrams and equations; *Martindale* in its latest edition includes structural formulae.

Pharmaceutics is one of the fundamental bases of pharmacy. Few, if any, other disciplines study the subject. Knowledge of the essentials which we have put down in this *FASTtrack* book are, in our view, very important if pharmacists are to continue to know about drugs and formulations and to contribute something special to healthcare. We hope that this book helps in preparing not only for examinations but also for the future.

David Attwood
Alexander T. Florence
July 2007

About the author

ALEXANDER FLORENCE recently retired as Dean of The School of Pharmacy at the University of London; he was previously James P. Todd Professor of Pharmaceutics at the University of Strathclyde. His research and teaching interests are drug delivery and targeting, dendrimers, nanoparticles, non-aqueous emulsions, novel solvents for use in pharmaceutics. He co-authored the book *Surfactant Systems: their Chemistry, Pharmacy and Biology* with David Attwood.

DAVID ATTWOOD is Professor of Pharmacy at the University of Manchester; he previously lectured at the University of Strathclyde. His research interests are in the physicochemical properties of drugs and surfactants, and in polymeric drug delivery systems. He has many years' experience in the teaching of physical pharmacy.

chapter 1
Solids

Overview

In this chapter we will:

- examine the various types of unit cell from which crystals are constructed and how the crystal lattice may be described using the system of Miller indices
- see how the external appearance of crystals may be described in terms of their habit and look at the various factors influencing the crystal habit
- discuss the formation of polymorphs and crystal hydrates by some drugs and examine the pharmaceutical consequences
- examine the process of wetting of solids and the importance of the contact angle in describing wettability
- look at the factors influencing the rate of dissolution of drugs and how drug solubility may be increased by forming eutectic mixtures.

Crystal structure and external appearance

- All crystals are constructed from repeating units called *unit cells.*
- All unit cells in a specific crystal are the same size and contain the same number of molecules or ions arranged in the same way.
- There are seven primitive unit cells (Figure 1.1): cubic, hexagonal, trigonal, tetragonal, orthorhombic, monoclinic and triclinic. Certain of these may also be end-centred (monoclinic and orthorhombic), body-centred (cubic, tetragonal and orthorhombic) or face-centred (cubic and orthorhombic), making a total of 14 possible unit cells called *Bravais lattices* (Figure 1.2).
- It is possible to describe the various planes of a crystal using the system of *Miller indices* (Figure 1.3). The general rules for applying this system are:
- Determine the intercepts of the plane on the *a*, *b*, and *c* axes in terms of unit cell lengths.
- Take the reciprocals of the intercepts.
- Clear any fractions by multiplying by the lowest common denominator.

KeyPoints

- The crystal lattice is constructed from repeating units called unit cells, of which there are only 14 possible types.
- The various planes of the crystal lattice can be described using the system of Miller indices.
- The external appearance of a crystal (crystal habit) depends on the conditions of crystallisation and affects the formulation properties of the crystal.

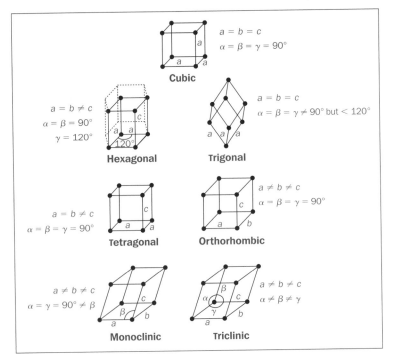

Fig. 1.1 The seven possible primitive unit cells.

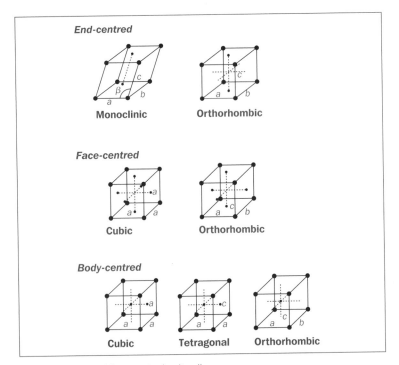

Fig. 1.2 End-centred, body-centred and face-centred unit cells.

- Reduce the numbers to the lowest terms.
- Replace negative numbers with a bar above the number.
- Express the result as three numbers, e.g. 101.

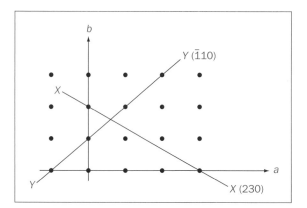

Figure 1.3 The Miller indices for two planes in a two-dimensional lattice.

The external appearance of a crystal is described by its overall shape or *habit*, for example, acicular (needle-like), prismatic or tabular. The crystal habit affects:

- the ability to inject a suspension containing a drug in crystal form – plate-like crystals are easier to inject through a fine needle than needle-like crystals
- the flow properties of the drug in the solid state – equidimensional crystals have better flow properties and compaction characteristics than needle-like crystals, making them more suitable for tableting.

The crystal habit depends on the conditions of crystallisation, such as solvent used, the temperature, and the concentration and presence of impurities. Surfactants in the solvent medium used for crystal growth can alter crystal form by adsorbing onto growing faces during crystal growth.

Polymorphism

When polymorphism occurs, the molecules arrange themselves in two or more different ways in the crystal; either they may be packed differently in the crystal lattice or there may be differences in the orientation or conformation of the molecules at the lattice sites.

Tips

Notice that:
- The smaller the number in the Miller index for a particular axis, the more parallel is the plane to that axis.
- A zero value indicates a plane exactly parallel to that axis.
- The larger a Miller index, the more nearly perpendicular a plane is to that axis.

Tip

The *habit* describes the overall shape of the crystal in general terms and so, for example, acicular crystals can have a large number of faces or can be very simple.

KeyPoints

- The crystals of some drugs can exist in more than one polymorphic form or as different solvates.
- Polymorphs and solvates of the same drug have different properties and this may cause problems in their formulation, analysis and absorption.

Polymorphs of the same drug have different X-ray diffraction patterns, may have different melting points and solubilities and also usually exist in different habits.

Certain classes of drug are particularly susceptible to polymorphism; for example, about 65% of the commercial sulfonamides and 70% of the barbiturates used medicinally are known to exist in several polymorphic forms.

The particular polymorph formed by a drug depends on the conditions of crystallisation; for example, the solvent used, the rate of crystallisation and the temperature.

Under a given set of conditions the polymorphic form with the lowest free energy will be the most stable, and other polymorphs will tend to transform into it.

Polymorphism has the following pharmaceutical implications:

Formulation problems

- Polymorphs with certain crystal habits may be difficult to inject in suspension form or to formulate as tablets (see above).
- Transformation between polymorphic forms during storage can cause changes in crystal size in suspensions and their eventual caking.
- Crystal growth in creams as a result of phase transformation can cause the cream to become gritty.
- Changes in polymorphic forms of vehicles such as theobroma oil, used to make suppositories, could cause products with different and unacceptable melting characteristics.

Analytical issues

- Difficulties in identification arise when samples that are thought to be the same substance give different infrared spectra in the solid state because they exist in different polymorphic forms.
- Change in polymorphic form can be caused by grinding with potassium bromide when samples are being prepared for infrared analysis.
- Changes in crystal form can also be induced by solvent extraction methods used for isolation of drugs from formulations prior to examination by infrared spectroscopy – these can be avoided by converting both samples and reference material into the same form by recrystallisation from the same solvent.

Tip

It is not possible to predict whether a particular drug will exist as several polymorphic forms and hence caution should always be applied when processing drugs to avoid possible changes in polymorphic form and hence of their properties.

Bioavailability differences

- The difference in the bioavailability of different polymorphic forms of a drug is usually insignificant but is a problem in the case of the chloramphenicol palmitate, one (form A) of the three polymorphic forms of which is poorly absorbed.

Crystal hydrates

When some compounds crystallise they may entrap solvent in the crystal. Crystals that contain solvent of crystallisation are called crystal *solvates*, or crystal *hydrates* when water is the solvent of crystallisation. Crystals that contain no water of crystallisation are termed *anhydrates*.

There are two main types of crystal solvate:

1. *Polymorphic solvates* are very stable and are difficult to desolvate because the solvent plays a key role in holding the crystal together. When these crystals lose their solvent they collapse and recrystallise in a new crystal form.
2. *Pseudopolymorphic solvates* lose their solvent more readily and desolvation does not destroy the crystal lattice. In these solvates the solvent is not part of the crystal bonding and merely occupies voids in the crystal.

The particular solvate formed by a drug depends on the conditions of crystallisation, particularly the solvent used.

The solvated forms of a drug have different physicochemical properties to the anhydrous form:

- The melting point of the anhydrous crystal is usually higher than that of the hydrate.
- Anhydrous crystals usually have higher aqueous solubilities than hydrates.
- The rates of dissolution of various solvated forms of a drug differ but are generally higher than that of the anhydrous form.
- There may be measurable differences in bioavailabilities of the solvates of a particular drug; for example, the monoethanol solvate of prednisolone tertiary butyl acetate has an absorption rate in vivo which is nearly five times greater than that of the anhydrous form of this drug.

KeyPoints

Wetting of solid surfaces and powders

The wetting of a solid when a liquid spreads over its surface is referred to as *spreading wetting*.

- The forces acting on a drop on the solid surface (Figure 1.4a) are represented by *Young's equation*:

$$\gamma_{S/A} = \gamma_{S/L} + \gamma_{L/A} \cos \theta$$

where $\gamma_{S/A}$ is the surface tension of the solid, $\gamma_{S/L}$ is the solid–liquid interfacial tension, $\gamma_{L/A}$ is the surface tension of the liquid and θ is the contact angle.

- The tendency for wetting is expressed by the spreading coefficient, S, as:

$$S = \gamma_{L/A} (\cos \theta - 1)$$

- For complete spreading of the liquid over the solid surface, S should have a zero or positive value.
- If the contact angle is larger than 0°, the term $(\cos \theta - 1)$ will be negative, as will the value of S.
- The condition for complete, spontaneous wetting is thus a zero value for the contact angle.

The wetting of a powder when it is initially immersed in a liquid is referred to as *immersional wetting* (once it has submerged, the process of spreading wetting becomes important).

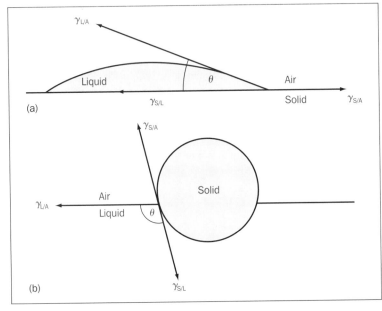

Figure 1.4 Equilibrium between forces acting on (a) a drop of liquid on a solid surface and (b) a partially immersed solid.

- The effectiveness of immersional wetting may be related to the contact angle which the solid makes with the liquid–air interface (Figure 1.4b).
- Contact angles of greater than 90° indicate wetting problems, for example when the drugs are formulated as suspensions.
- Examples of very hydrophobic (non-wetting) drugs include magnesium and aluminium stearates, salicylic acid, phenylbutazone and chloramphenicol palmitate.
- The normal method of improving wettability is by the inclusion of surfactants in the formulation. The surfactants not only reduce $\gamma_{L/A}$ but also adsorb on to the surface of the powder, thus reducing $\gamma_{S/L}$. Both of these effects reduce the contact angle and improve the dispersibility of the powder.

Dissolution of drugs

The rate of dissolution of solids is described by the *Noyes–Whitney* equation:

$$\frac{dw}{dt} = \frac{DA}{\delta}(c_s - c)$$

where dw/dt is the rate of increase of the amount of material in solution dissolving from a solid; c_s is the saturation solubility of the drug in solution in the diffusion layer and c is the concentration of drug in the bulk solution, A is the area of the solvate particles exposed to the solvent, δ is the thickness of the diffusion layer and D is the diffusion coefficient of the dissolved solute. This equation predicts:

- a decrease of dissolution rate because of a decrease of D when the viscosity of the medium is increased
- an increase of dissolution rate if the particle size is reduced by micronisation because of an increase in A
- an increase of dissolution rate by agitation in the gut or in a flask because of a decrease in δ
- an increase of dissolution rate when the concentration of drug is decreased by intake of fluid, and by removal of drug by partition or absorption
- a change of dissolution rate when c_s is changed by alteration of pH (if the drug is a weak electrolyte).

Solid dispersions

A solid solution comprises a solid solute molecularly dispersed in a solid solvent and is designed to improve the biopharmaceutical properties of drugs that are poorly soluble or difficult to wet.

Solid dispersions are *eutectic* mixtures comprising drug in microcrystalline form and a substance that is readily soluble in water (a carrier).

Below the eutectic temperature the mixture consists of a microcrystalline mixture of the components (Figure 1.5).

When the solid dispersion is added to water the soluble carrier dissolves, leaving the drug dispersed as very fine crystals which dissolve rapidly.

Examples of solid dispersions include griseofulvin–succinic acid, chloramphenicol–urea, sulfathiazole–urea, and niacinamide–ascorbic acid.

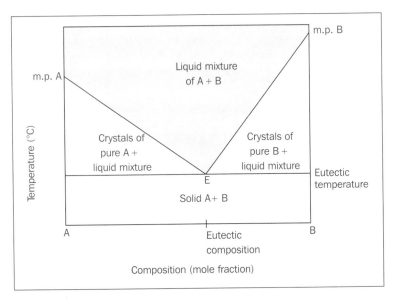

Figure 1.5 Phase diagram showing eutectic point, E.

Multiple choice questions

1. **Which unit cell is characterised by the following parameters?**

$$\begin{cases} a \neq b \neq c \\ \alpha = \gamma = 90^{\circ} \\ \beta \neq 90^{\circ} \end{cases}$$

   a. triclinic
   b. monoclinic
   c. orthorhombic
   d. tetragonal
   e. hexagonal

2. From the list below, select which of the statements may be correctly applied to an orthorhombic unit cell:
a. The angles $\alpha = \beta = \gamma = 90°$
b. The angles $\alpha \neq \beta \neq \gamma \neq 90°$
c. The angles $\alpha = \beta = 90°$; $\gamma = 120°$
d. The lengths $a \neq b \neq c$
e. The lengths $a = b \neq c$

3. The Miller indices of a plane which intersects the a, b and c axes at $a = 3$, $b = 2$ and $c = \infty$ are:
a. 320
b. 460
c. 230
d. $\bar{2}30$
e. $\bar{3}20$

4. From the list below, select which of the statements may be correctly applied to a pseudopolymorphic solvate:
a. The solvent is strongly bound.
b. Desolvation does not destroy the crystal lattice.
c. Desolvation leads to collapse and recrystallisation in a different form.
d. The solvent is easily removed.
e. The solvent plays a key role in holding together the crystal lattice.

5. From the list below select which of the statements may be correctly applied to the wetting of a flat solid surface by a liquid:
a. The wetting is referred to as spreading wetting.
b. Complete, spontaneous wetting of the surface will occur when the contact angle is greater than 90°.
c. The condition for spontaneous wetting is a zero contact angle.
d. Spreading occurs more readily if the surface tension of the liquid is high.
e. Spreading will occur if the spreading coefficient is negative.

6. From the list below select which of the statements may be correctly applied to the wetting of a powder when immersed in a liquid:
a. Complete, spontaneous wetting of the powder will occur when the contact angle is greater than 90°.
b. Wetting will only occur if the contact angle is zero.
c. The wetting is referred to as spreading wetting.
d. Wettability may be improved by reducing the surface tension of the liquid.
e. Wettability may be improved by reducing the hydrophobicity of the solid surface.

7. **Indicate whether each of the following statements is *true* or *false*. The Noyes–Whitney equation predicts an increase of dissolution rate when:**
a. The viscosity of the medium is increased.
b. The particle size is reduced.
c. The liquid medium is agitated.
d. The saturated solubility of solid is decreased.
e. Dissolved drug is removed from solution.

8. **Two components A and B (the melting point of component A is lower than that of component B) form solid dispersions. Indicate which of the following statements is *true* or *false*.**
a. Below the eutectic temperature the system consists of microcrystals of A dissolved in liquid B.
b. Below the eutectic temperature the system consists of microcrystals of B dissolved in liquid A.
c. Below the eutectic temperature the system consists of a mixture of microcrystals of A and B in solid form.
d. On cooling a solution of A and B which is richer in A than B crystals of B will appear.
e. The eutectic mixture has the lowest melting point of any mixture.

chapter 2
Solubility and solution properties of drugs

Overview

In this chapter we will:

- examine the properties of liquefied gases (propellants) used as solvents for drugs delivered in aerosol devices and in particular consider the factors influencing the vapour pressure in these devices
- consider the factors controlling the solubility of drugs in solution, in particular the nature of the drug molecule and the crystalline form in which it exists, its hydrophobicity, its shape, its surface area, its state of ionisation, the influence of pH of the medium and the importance of the pK_a of the drug
- look at the effect of pH on the ionisation of drugs in aqueous solution. See how to calculate the pH of solutions of drugs from a knowledge of their pK_a and how to prepare buffer solutions to control pH
- consider some thermodynamic properties of drugs in solution such as activity and chemical potential
- examine the partitioning of drugs between two immiscible phases and their diffusional properties in solution.

Solvents for pharmaceutical aerosols

Liquified gases under pressure in aerosol devices revert to the gaseous state when the device is activated and the liquid reaches atmospheric pressure. The drug is suspended or dissolved in the liquefied gas (propellant) and the drug–propellant mixture is expelled when the device is activated.

Hydrofluoroalkanes (HFAs) now replace chlorofluorocarbon (CFC) propellants in pressurised metered-dose inhalers because of the ozone-depleting properties of CFCs (Montreal Protocol 1989).

There is an equilibrium between a liquefied propellant and its vapour and there is a *vapour pressure* above the liquid, the value of which is determined by the propellants used and the presence of dissolved solutes.

The vapour pressure above the aerosol mixture determines the aerosol droplet size. In metered-dose inhalers, for example, this has an important influence on the efficiency of deposition in the lungs.

KeyPoints

- A *solution* can be defined as a system in which molecules of a solute (such as a drug or protein) are dissolved in a solvent vehicle.
- If the solvent is volatile there will be a vapour pressure above the solution which depends on the solvent's properties, the presence of solute and the temperature.
- When a solution contains a solute at the limit of its solubility at any given temperature and pressure it is said to be *saturated*.
- If the solubility limit is exceeded, solid particles of solute may be present and the solution phase will be in equilibrium with the solid, although under certain circumstances *supersaturated solutions* may be prepared, where the drug exists in solution above its normal solubility limit.
- The maximum *equilibrium solubility* of a drug dictates the *rate of solution (dissolution)* of the drug; the higher the solubility, the more rapid is the rate of solution.

The vapour pressure above a liquid propellant is constant at constant temperature and this property is exploited in the design of the totally implantable infusion pump (Figure 2.1):

- This device is implanted under the skin in the lower abdomen.
- It delivers infusate containing the appropriate drug at a constant rate (usually 1 cm^3 day^{-1}) into an artery or vein.
- It consists of a relatively small (9 × 2.5 cm) titanium disc which is divided into two chambers by cylindrical titanium bellows that form a flexible but impermeable barrier between the compartments.
- The outer compartment contains freon; the inner compartment contains the infusate and connects via a catheter to a vein or artery through a series of filters and flow-regulating resistant elements.
- The vapour pressure above the liquid propellant remains constant because of the relatively constant temperature of the body, and hence a constant pressure is exerted on the bellows, ensuring a constant rate of delivery of infusate into the blood stream.
- The propellant is replenished as required by a simple percutaneous injection through the skin.

Raoult's law is important because it allows the calculation of vapour pressure above an aerosol mixture from a knowledge of the composition of the solution:

- It gives the relationship between the partial vapour pressure of a component i in the vapour phase, p_i, and the mole fraction of that component in solution, x_i, (assuming ideal behaviour) as:

$$p_i = p_i^{\ominus} x_i$$

where $p_i^{\ominus}$ is the vapour pressure of the pure component.
- Binary mixtures of hydrofluoroalkanes show behaviour which approaches ideality.
- There may be large positive deviations from Raoult's law when cosolvents such as alcohol are included in the aerosol formulation to enhance its solvent power.

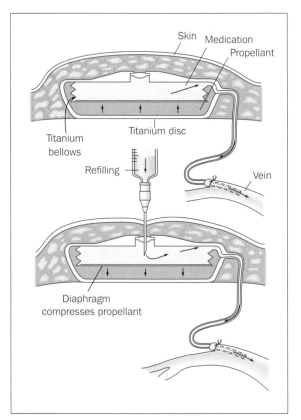

Figure 2.1 Diagram of the Infusaid implantable infusion pump during operation (top) and during refilling (bottom). Reproduced from P.J. Blackshear and T.H. Rhode. In *Controlled Drug Delivery*, Vol 2. *Clinical Applications*. Ed. S.D. Burk. Boca Raton, FL, CRC Press, p11.

Tips

- Raoult's law can be used to calculate the lowering of vapour pressure following the addition of a non-volatile solute to a solvent. The equation is,

$$\frac{p_1^{\ominus} - p_1}{p_1^{\ominus}} = x_2$$

which shows that the relative lowering of the vapour pressure is equal to the mole fraction x_2 of the solute.

When calculating the total vapour pressure above an aerosol mixture you will need to:
- Apply Raoult's law to calculate the partial pressures of each component from the mole fraction of each in the mixture.
- Calculate the total vapour pressure P using Dalton's law of partial pressures, which states that P is the sum of the partial pressures of the component gases, assuming ideal behaviour.
- Convert the pressure (if required) from Pa to pounds per square inch gauge (psig) using 1 Pa = (1/6894.76) − 14.7 psig.

Factors influencing solubility

The solution process can be considered in three stages
(Figure 2.2):

1. A solute (drug) molecule is 'removed' from its crystal.
2. A cavity for the molecule is created in the solvent.
3. The solute molecule is inserted into this cavity.

- The *surface area* of the drug molecule affects solubility
  because placing the solute molecule in the solvent cavity
  (step 3) requires a number of solute–solvent contacts; the
  larger the solute molecule, the larger the cavity required (step
  2) and the greater the number of contacts created. For simple
  molecules solubility decreases with increase of molecular
  surface area.

- The *boiling point* of liquids and the *melting point* of solids
  both reflect the strengths of interactions between the
  molecules in the pure liquid or the solid state (step 1). In
  general, aqueous solubility decreases with increasing boiling
  point and melting point.

Figure 2.2 Diagrammatic representation of the processes involved in the dissolution of a crystalline solute.

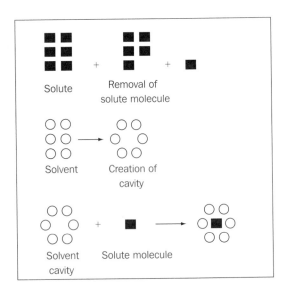

- The influence of *substituents* on the solubility of molecules
  in water can be due to their effect on the properties of the
  solid or liquid (for example, on its molecular cohesion, step
  1) or to the effect of the substituent on its interaction with
  water molecules (step 3). Substituents can be classified
  as either hydrophobic or hydrophilic, depending on their
  polarity:

  – Polar groups such as –OH capable of hydrogen bonding with
    water molecules impart high solubility.

- Non-polar groups such as $-CH_3$ and $-Cl$ are hydrophobic and impart low solubility.
- Ionisation of the substituent increases solubility, e.g. $-COOH$ and $-NH_2$ are slightly hydrophilic whereas $-COO^-$ and $-NH_3^+$ are very hydrophilic.
- The position of the substituent on the molecule can influence its effect on solubility, for example the aqueous solubilities of *o*-, *m*- and *p*-dihydroxybenzenes are 4, 9 and 0.6 mol dm^{-3}, respectively.
- The solubility of *inorganic electrolytes* is influenced by their crystal properties and the interaction of their ions with water (hydration). If the heat of hydration is sufficient to provide the energy needed to overcome the lattice forces, the salt will be freely soluble at a given temperature and the ions will readily dislodge from the crystal lattice.
- *Additives* may either increase or decrease the solubility of a solute in a given solvent; their effect on the solubility of sparingly soluble solutes may be evaluated using the *solubility product*. Salts that increase solubility are said to *salt in* the solute and those that decrease solubility *salt out* the solute. The effect that they have will depend on several factors:
- the effect the additive has on the structure of water
- the interaction of the additive with the solute
- the interaction of the additive with the solvent.

pH is one of the primary influences on the solubility of most drugs that contain ionisable groups (Figure 2.3):
- *Acidic drugs*, such as the non-steroidal anti-inflammatory agents, are less soluble in acidic solutions than in alkaline solutions because the predominant undissociated species cannot interact with water molecules to the same extent as the ionised form which is readily hydrated. The equation relating the solubility, S, of an acidic drug to the pH of the solution is:

$$pH - pK_a = \log\left(\frac{S - S_o}{S_o}\right)$$

where S_o is the solubility of the undissociated form of the drug.
- *Basic drugs* such as ranitidine are more soluble in acidic solutions where the ionised form of the drug is predominant. The equation relating the solubility, S, of a basic drug to the pH of the solution is:

$$pH - pK_a = \log\left(\frac{S_o}{S - S_o}\right)$$

Tip

The *solubility product*, K_{sp}, of a sparingly soluble solute such as silver chloride is written as:

$$K_{sp} = [Ag^+][Cl^-]$$

Therefore, if either $[Ag^+]$ or $[Cl^-]$ is increased by adding an Ag^+ or Cl^- ion to the solution then because the value of the solubility constant cannot change, some of the sparingly soluble salt will precipitate, i.e. the solubility of the sparingly soluble is decreased by adding a common ion (referred to as the *common ion effect*).

Tip

At a particular pH, known as the *isoelectric point*, pH$_i$, the effective net charge on the amphoteric molecule is zero. pH$_i$ can be calculated from:

$$pH_i = (pK_a^{acidic} + pK_a^{basic})/2$$

where pK$_a^{acidic}$ and pK$_a^{basic}$ are the pK$_a$s of the acidic and basic groups respectively.

Tip

As a rough guide, the solubility of drugs in which the unionised species has a low solubility varies by a factor of 10 for each pH unit change.

KeyPoints

- Many drugs are weak organic acids (for example, acetylsalicylic acid) or weak organic bases (for example, procaine) or their salts (for example, ephedrine hydrochloride).
- A *weak* acid or base is only *slightly ionised* in solution, unlike a *strong* acid or base, which is *completely ionised*.
- The degree to which weak acids and bases are ionised in solution is highly dependent on the pH.
- The exceptions to this general statement are the non-electrolytes, such as the steroids, and the quaternary ammonium compounds, which are completely ionised at all pH values and in this respect behave as strong electrolytes.
- The extent of ionisation of a drug has an important effect on its absorption, distribution and elimination.

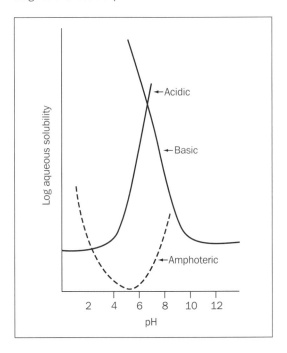

Figure 2.3 Solubility of acidic, basic and amphoteric drugs as a function of pH.

- *Amphoteric drugs* such as the sulfonamides and tetracyclines display both basic and acidic characteristics. The zwitterion has the lowest solubility, S_o, and the variation of solubility with pH is given by:

$$pH - pK_a = \log\left(\frac{S_o}{S - S_o}\right)$$

at pH values below the isoelectric point and

$$pH - pK_a = \log\left(\frac{S - S_o}{S_o}\right)$$

at pH values above the isoelectric point.

Ionisation of drugs in solution

Ionisation of weakly acidic drugs and their salts

- If the weak acid is represented by HA, its ionisation in water may be represented by the equilibrium:

$$HA + H_2O \leftrightharpoons A^- + H_3O^+$$

- The equilibrium constant, K_a, is referred to as the *ionisation constant, dissociation constant* or *acidity constant* and is given by:

$$K_a = \frac{[H_3O^+][A^-]}{[HA]}$$

- The negative logarithm of K_a is referred to as pK_a, i.e. $pK_a = -\log K_a$.
- When the pH of an aqueous solution of the weakly acidic drug approaches to within 2 pH units of the pK_a there is a very pronounced change in the ionisation of that drug (Figure 2.4):

Figure 2.4 Percentage ionisation of weakly acidic and weakly basic drugs as a function of pH.

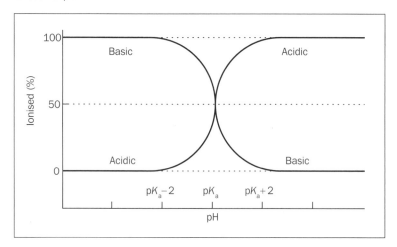

- The percentage ionisation at a given pH can be calculated from:

$$\text{percentage ionisation} = \frac{100}{1 + \text{antilog}\,(pK_a - pH)}$$

- Weakly acidic drugs are virtually completely unionised at pHs up to 2 units below their pK_a and virtually completely ionised at pHs greater than 2 units above their pK_a. They are exactly 50% ionised at their pK_a values.
- Salts of weak acids are essentially completely ionised in solution, for example when sodium salicylate (salt of the weak acid, salicylic acid, and the strong base NaOH) is dissolved in water, it ionises almost entirely into the conjugate base of salicylic acid, $HOC_6H_5COO^-$, and Na^+ ions. The conjugate acids formed in this way are subject to acid–base equilibria described by the general equations above.

Ionisation of weakly basic drugs and their salts

- If the weak acid is represented by B, its ionisation in water may be represented by the equilibrium:

  $B + H_2O \leftrightarrows BH^+ + OH^-$

- The equilibrium constant, K_b, is referred to as the *ionisation constant, dissociation constant* or *basicity constant* and is given by:

  $$K_b = \frac{[OH^-][BH^+]}{[B]}$$

- The negative logarithm of K_b is referred to as pK_b, i.e. $pK_b = -\log K_b$.

- The percentage ionisation at a given pH can be calculated from:

  $$\text{percentage ionisation} = \frac{100}{1 + \text{antilog}\,(pH - pK_w + pK_b)}$$

- Weakly basic drugs are virtually completely ionised at pHs up to 2 units below their pK_a and virtually completely unionised at pHs greater than 2 units above their pK_a. They are exactly 50% ionised at pHs equal to their pK_a values.

- Salts of weak bases are essentially completely ionised in solution; for example, ephedrine hydrochloride (salt of the weak base, ephedrine, and the strong acid HCl) exists in aqueous solution in the form of the conjugate acid of the weak base, $C_6H_5CH(OH)CH(CH_3)N^+H_2CH_3$, together with its Cl$^-$ counterions. The conjugate bases formed in this way are subject to acid–base equilibria described by the general equations above.

Ionisation of amphoteric drugs

- These can function as either weak acids or weak bases in aqueous solution depending on the pH and have pK_a values corresponding to the ionisation of each group.

- If pK_a of the acidic group, pK_a^{acidic}, is higher than that of the basic group, pK_a^{basic}, they are referred to as ordinary ampholytes and exist in solution as a cation, an unionised form, and an anion depending on the pH of the solution. For example, the ionisation of *m*-aminophenol ($pK_a^{\text{acidic}} = 9.8$ and $pK_a^{\text{basic}} = 4.4$) changes as the pH increases as follows:

  $NH_3^+C_6H_4OH \leftrightarrows NH_2C_6H_4OH \leftrightarrows NH_2C_6H_5O^-$

Tips

- It is usual to use only pK_a values when referring to both weak acids and bases.
- pK_a and pK_b values of conjugate acid–base pairs are linked by the expression:

  $pK_a + pK_b = pK_w$

  where pK_w is the negative logarithm of the dissociation constant for water, K_w.
- $pK_w = 14.00$ at 25°C (but decreases with temperature increase) and hence a value of pK_b can easily be calculated if required.
- pK_a and pK_b values provide a convenient means of comparing the strengths of weak acids and bases. The lower the pK_a, the stronger the acid; the lower the pK_b, the stronger is the base (but note they are still weak acids or bases).

- If $pK_a^{acidic} < pK_a^{basic}$ they are referred to as zwitterionic ampholytes and exist in solution as a cation, a zwitterion (having both positive and negative charges), and an anion depending on the pH of the solution. Examples of this type of compound include the amino acids, peptides and proteins. Glycine ($pK_a^{acidic} = 2.3$ and $pK_a^{basic} = 9.6$) ionises as follows:

  $HOOC\ CH_2\ NH_3^+ \leftrightharpoons {}^-OOC\ CH_2\ NH_3^+ \leftrightharpoons {}^-OOC\ CH_2\ NH_2$

- The ionisation pattern of both types is more complex, however, with drugs in which the difference in pK_a of the two groups is much smaller (< 2 pH units) because of overlapping of the two equilibria.

Ionisation of polyprotic drugs

- Several acids, for example citric, phosphoric and tartaric acid, are capable of donating more than one proton and these compounds are referred to as *polyprotic or polybasic acids*. Similarly, polyprotic bases are capable of accepting two or more protons. Examples of polyprotic drugs include the polybasic acids amoxicillin and fluorouracil, and the polyacidic bases pilocarpine, doxorubicin and aciclovir.

- Each stage of the dissociation of the drug may be represented by an equilibrium expression and hence each stage has a distinct pK_a or pK_b value. For example, the ionisation of phosphoric acid occurs in three stages:

$$H_3PO_4 + H_2O \leftrightharpoons H_2PO_4^- + H_3O^+ \qquad pK_1 = 2.1$$
$$H_2PO_4^- + H_2O \leftrightharpoons HPO_4^{2-} + H_3O^+ \qquad pK_2 = 7.2$$
$$HPO_4^{2-} + H_2O \leftrightharpoons PO_4^{3-} + H_3O^+ \qquad pK_3 = 12.7$$

- If the pK_a values of each stage of dissociation are far apart it is usually possible to assign them to the ionisation of specific groups, but if they are within about 2 pH units of each other this is not possible. For a more complete picture of the dissociation it is necessary to take into account all possible ways in which the molecule may be ionised and all the possible species present in solution. In this case the constants are called *microdissociation constants*.

pH of drug solutions

The pH of a *strong* acid such as HCl is given by $pH = -\log[H^+]$. This is because strong acids are completely ionised in solution. However, as seen above, *weak* acids and bases are only slightly ionised in solution and the extent of their ionisation changes with pH and so therefore does their pH. The pH at any particular concentration, c, can be calculated from the pK_a value:

- *Weakly acidic drugs:* $pH = \frac{1}{2}\ pK_a - \frac{1}{2}\log c.$
- *Weakly basic drugs:* $pH = \frac{1}{2}\ pK_w + \frac{1}{2}\ pK_a + \frac{1}{2}\log c.$

Tips

- *Drug salts*:
  - Salts of a weak acid and a strong base:
  $$pH = \tfrac{1}{2}\,pK_w + \tfrac{1}{2}\,pK_a + \tfrac{1}{2}\log c$$
  - Salts of a weak base and a strong acid:
  $$pH = \tfrac{1}{2}\,pK_a - \tfrac{1}{2}\log c$$
  - Salts of a weak acid and a weak base:
  $$pH = \tfrac{1}{2}\,pK_w + \tfrac{1}{2}\,pK_a - \tfrac{1}{2}\,pK_b$$
  (note that there is no concentration term in this equation, meaning that the pH does not vary with concentration).

Buffers

- Buffers are usually mixtures of a weak acid and its salt (that is, a conjugate base), or a weak base and its conjugate acid.
- A mixture of a weak acid HA and its ionised salt (for example, NaA) acts as a buffer because the A^- ions from the salt combine with the added H^+ ions, removing them from solution as undissociated weak acid:

$$A^- + H_3O^+ \leftrightarrows H_2O + HA$$

Added OH^- ions are removed by combination with the weak acid to form undissociated water molecules:

$$HA + OH^- \leftrightarrows H_2O + A^-$$

- A mixture of a weak base and its salt acts as a buffer because added H^+ ions are removed by the base B to form the salt and OH^- ions are removed by the salt to form undissociated water:
$$B + H_3O^+ \leftrightarrows H_2O + BH^+$$
$$BH^+ + OH^- \leftrightarrows H_2O + B$$
- The concentration of buffer components required to maintain a solution at the required pH may be calculated from the *Henderson–Hasselbalch equations*:

Weak acid and its salt:
$$pH = pK_a + \log\frac{[salt]}{[acid]}$$

Weak base and its salt:
$$pH = pK_w - pK_b + \log\frac{[base]}{[salt]}$$

- The effectiveness of a buffer in minimising pH change is expressed as the buffer capacity, β, calculated from:

$$\beta = \frac{2.303\, c_0\, K_a\, [H_3O^+]}{([\,H_3O^+]+K_a)^2}$$

where c_0 is the total initial buffer concentration.

A plot of β against pH (Figure 2.5) shows that:

- The buffer capacity is maximum when pH = pK_a.
- Maximum buffer capacity, β_{max}, = $0.576 c_0$.
- If, instead of using a single weak monobasic acid, which has a maximum buffer capacity at pH = pK_a, you use a suitable mixture of polybasic and monobasic acids, it is possible to produce a buffer which is effective over a wide pH range because each stage of the ionisation of the polybasic acid has its own β_{max} value. Such solutions are referred to as *universal buffers*. A typical example is a mixture of citric acid (pK_{a1} = 3.06, pK_{a2} = 4.78 and pK_{a3} = 5.40), Na_2HPO_4(pK_a of conjugate acid, $H_2PO_4^-$ = 7.2), diethylbarbituric acid (pK_{a1} = 7.43) and boric acid (pK_{a1} = 9.24). This buffer is effective over a pH range 2.4 to 12.

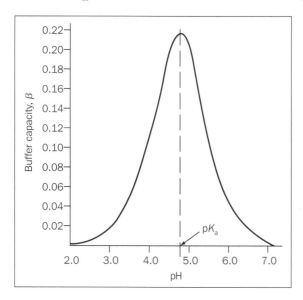

Figure 2.5 Buffer capacity of a weak acid/salt buffer as a function of pH, showing maximum buffer capacity at pK_a.

Thermodynamic properties of drugs in solution

Activity and activity coefficient

- In any real solution interactions occur between the components which reduce the *effective* concentration of the solution. The *activity* is a way of describing this effective concentration.
- The ratio of the activity to the concentration is called the *activity coefficient*, γ, that is, γ = activity/concentration. Therefore, for an ideal solution $\gamma = 1$.

Tips

- Salts such as ephedrine hydrochloride ($C_6H_5CH(OH)CH(NHCH_3)CH_3HCl$) are 1:1 (or uni-univalent) electrolytes; that is, on dissociation each mole yields one cation $C_6H_5CH(OH)CH(N^+H_2CH_3)CH_3$, and one anion, Cl^-.
- Other salts are more complex in their ionisation behaviour; for example, ephedrine sulfate is a 1:2 electrolyte, each mole giving two moles of the cation and one mole of SO_4^{2-} ions.

- When drugs are salts they ionise in solution and the activity of each ion is the product of its activity coefficient and its molar concentration, that is, $a_+ = \gamma_+ m_+$ and $a_- = \gamma_- m_-$. However, because it is not possible to measure ionic activities separately we use mean ionic parameters, i.e. $\gamma_\pm = a_\pm/m_\pm$.
- The mean ion activity coefficient $\gamma_\pm$ can be calculated using the Debye–Hückel equation:
$$-\log \gamma_\pm = z_+ z_- A\sqrt{I}$$
where z_+ and z_- are the valencies of the ions, A is a constant ($A = 0.509$ in water at 298 K) and I is the total ionic strength. For a 1:1 electrolyte, $I = m$, for a 1:2 electrolyte $I = 3m$ and for a 2:2 electrolyte, $I = 4m$.

Chemical potential

- Chemical potential is the *effective* free energy per mole of each component in the mixture and is always less than the free energy of the pure substance.
- The chemical potential of a component in a two-phase system (for example, oil and water) at equilibrium at a fixed temperature and pressure is identical in both phases. A substance in a two-phase system which is not at equilibrium will have a tendency to diffuse spontaneously from a phase in which it has a high chemical potential to another in which it has a low chemical potential. The difference in chemical potential is the driving force for diffusion between the two phases.
- The chemical potential μ_2 of a non-ionised component in dilute solution is given by:
$$\mu_2 = \mu_2^\ominus + RT \ln M_1 - RT \ln 1000 + RT \ln m$$
where $\mu_2^\ominus$ is the chemical potential of the component in its standard state and M_1 = molecular weight of the solvent.
- The chemical potential of a 1:1 electrolyte is given by:
$$\mu_2 = \mu_2^\ominus + 2RT \ln m\gamma_\pm.$$

Osmotic properties of drugs in solution – isotonic solutions

- Whenever a solution is separated from a solvent by a membrane that is only permeable to solvent molecules (referred to as a *semipermeable membrane*), there is a passage of solvent across the membrane into the solution. This is the phenomenon of *osmosis*.

- Solvent passes through the membrane because the chemical potentials on either side of the membrane are not equal. Since the chemical potential of a solvent molecule in solution is less than that in pure solvent, solvent will spontaneously enter the solution until the chemical potentials are the same.
- If the solution is totally confined by a semipermeable membrane and immersed in the solvent, then a pressure differential develops across the membrane: this is referred to as the *osmotic pressure*.
- The equation which relates the osmotic pressure of the solution, Π, to the solution concentration is the *van't Hoff equation*:

$$\Pi V = n_2 RT$$

where V is the molar volume of the solute and n_2 is the number of moles of solute.
- Osmotic pressure is a *colligative* property, which means that its value depends on the number of ions in solution. Therefore, if the drug is ionised you need to include the contribution of the counterions to the total number of ions in solution. Note, however, that the extent of ionisation changes as the solution is diluted and is only complete in very dilute solution.
- Because the red bood cell membrane acts as a semipermeable membrane it is important to ensure that the osmotic pressure of solutions for injection is approximately the same as that of blood serum. Such solutions are said to be *isotonic* with blood. Solutions with a higher osmotic pressure are *hypertonic* and those with a lower osmotic pressure are termed *hypotonic* solutions. Similarly, in order to avoid discomfort on administration of solutions to the delicate membranes of the body, such as the eyes, these solutions are made isotonic with the relevant tissues.
- Osmotic pressure is not a readily measurable quantity and it is usual to use the freezing-point depression (which is also a colligative property) when calculating the quantities required to make a solution isotonic. A solution which is isotonic with blood has a freezing-point depression, ΔT_f, of 0.52°C. Therefore the freezing point of the drug solution has to be adjusted to this value by adding sodium chloride to make the solution isotonic. The amount of the adjusting substance required can be calculated from:

$$w = \frac{0.52 - a}{b}$$

> **KeyPoint**
>
> Parenteral solutions should be of approximately the same tonicity as blood serum; the amount of adjusting substance which must be added to a formulation to achieve isotonicity can be calculated using the freezing point depressions of the drug and the adjusting substance.

> **Tip**
>
> Note that, although the freezing point of blood serum is –0.52°C, the freezing-point depression is +0.52°C because the word 'depression' implies a decrease in value. This is a common source of error in isotonicity calculations.

where w is the weight in grams of adjusting substance to be added to 100 cm^3 of drug solution to achieve isotonicity, a is the number of grams of drug in 100 ml of solution multiplied by ΔT_f of a 1% drug solution, and b is ΔT_f of 1% adjusting substance.

Partitioning of drugs between immiscible solvents

- Examples of partitioning include:
- – drugs partitioning between aqueous phases and lipid biophases.
- – preservative molecules in emulsions partitioning between the aqueous and oil phases.
- – antibiotics partitioning into microorganisms.
- – drugs and preservative molecules partitioning into the plastic of containers or giving sets.
- If two immiscible phases are placed in contact, one containing a solute soluble to some extent in both phases, the solute will distribute itself until the chemical potential of the solute in one phase is equal to its chemical potential in the other phase.
- The distribution of the solute between the two phases is represented by the *partition coefficient* or *distribution coefficient*, P, defined as the ratio of the solubility in the non-aqueous (oily) phase, C_o, to that in the aqueous phase, C_w, i.e.
$$P = C_o/C_w.$$
- It is usual to express the partitioning as log P. The greater the value of log P, the higher the lipid solubility of the solute.
- Octanol is usually used as the non-aqueous phase in experiments to measure the partition coefficient of drugs. Other non-aqueous solvents, for example isobutanol and hexane, have also been used.
- In many systems the ionisation of the solute in one or both phases or the association of the solute in one of the solvents complicates the calculation of partition coefficient:
- – For example, if the solute *associates* to form dimers in phase 2 then $K = \sqrt{C_2}/C_1$, where K is a constant combining the partition coefficient and the association constant and C_1 is the concentration in phase 1.
- – Many drugs will ionise in at least one phase, usually the aqueous phase. It is generally accepted that only the non-ionised species partitions from the aqueous phase into the non-aqueous phase. Ionised species, being hydrated and highly soluble in the aqueous phase, disfavour the organic phase because transfer of such a hydrated species involves dehydration. In addition, organic solvents of low polarity do not favour the existence of free ions.

- If ionisation and its consequences are neglected, an *apparent* partition coefficient, P_{app}, is obtained simply by assay of both phases, which will provide information on how much of the drug is present in each phase, regardless of status. The relationship between the true thermodynamic P and P_{app} is given by the following equations:
  - For *acidic* drugs:

$$\log P = \log P_{app} - \log\left(\frac{1}{1+10^{\,pH-pK_a}}\right)$$

  - For *basic* drugs:

$$\log P = \log P_{app} - \log\left(\frac{1}{1+10^{\,pK_a-pH}}\right)$$

  - For *amphoteric* compounds such as the tetracyclines, the pH dependence of partition coefficient is more complex than for most drugs, as the tetracyclines are amphoteric. For slightly simpler amphoteric compounds, such as *p*-aminobenzoic acid and sulfonamides, the apparent partition coefficient is maximal at the isoelectric point.
- Correlations between partition coefficients and biological activity are expressed by *Ferguson's principle* which states that, within reasonable limits, substances present at approximately the same proportional saturation (that is, with the same thermodynamic activity) in a given medium have the same biological potency.
- Other examples of partitioning of pharmaceutical importance include the permeation of antimicrobial agents into rubber stoppers and other closures, the partitioning of glyceryl trinitrate (volatile drug with a chloroform/water partition coefficient of 109) from simple tablet bases into the walls of plastic bottles and into plastic liners used in packaging tablets, and the permeation of drugs into polyvinyl chloride infusion bags.

Diffusion of drugs in solution

Drug molecules in solution will spontaneously diffuse from a region of high chemical potential to one of low chemical potential. Although the driving force for diffusion is the gradient of chemical potential, it is more usual to think of the diffusion process in terms of the concentration gradient.

The rate of diffusion may be calculated from *Fick's first law*:

$$J = -D\,(dc/dx)$$

where J is the flux of a component across a plane of unit area, dc/dx is the concentration gradient and D is the diffusion coefficient (or diffusivity). The negative sign indicates that the flux is in the direction of decreasing concentration. J is in mol m^{-2} s^{-1}, c is in mol m^{-3} and x is in m; therefore, the units of D are m^2 s^{-1}.

The relationship between the radius, a, of the diffusing molecule and its diffusion coefficient (assuming spherical particles or molecules) is given by the *Stokes–Einstein equation* as:

$$D = \frac{RT}{6\pi\eta a N_A}$$

where η is the viscosity of the drug solution and N_A is Avogadro's constant.

The diffusional properties of a drug have relevance in pharmaceutical systems in a consideration of such processes as the dissolution of the drug and transport through artificial (e.g. polymer) or biological membranes, and diffusion in tissues such as the skin or in tumours.

Multiple choice questions

1. **Calculate the vapour pressure in Pa and psig above an aerosol mixture consisting of 30% w/w of a propellant (molecular weight 170.9) with a vapour pressure of 1.90 × 10⁵ Pa and 70% w/w of a second propellant (molecular weight 120.9) with a vapour pressure of 5.85 × 10⁵ Pa. Assume ideal behaviour.**
   a. 4.67×10^5 Pa
   b. 4.49×10^5 Pa
   c. 7.75×10^5 Pa
   d. 50.42 psig
   e. 65.12 psig

2. **Indicate which of the following molecular characteristics will be expected to increase the solubility of a simple solute in an aqueous solution:**
   a. a low melting point
   b. the presence of a polar group
   c. a high molecular surface area
   d. the presence of an ionised group
   e. a high boiling point

3. **Indicate which of the following general statements are true:**
   a. Acidic drugs are less soluble in acidic solutions than in alkaline solutions.
   b. Basic drugs are more soluble in alkaline solutions than in acid solutions.
   c. The zwitterion of an amphoteric drug has a higher solubility than the acidic or basic forms of the drug.
   d. The effective net charge on the zwitterion is zero at the isoelectric point.

4. **Indicate which of the following general statements are true:**
   a. A weakly acidic drug is unionised when the pH of the solution is at least 2 pH units below its pK_a.

b. A weakly basic drug is fully ionised when the pH of the solution is at least 2 pH units greater than its pK_a.

c. Quaternary ammonium compounds are fully ionised at all pHs.

d. Weakly acidic drugs are 50% ionised when the pH of the solution is equal to their pK_a.

e. Salts of weak acids are fully ionised in solution.

f. The higher the pK_a of a weak acid, the stronger is the acid.

g. The sum of pK_a and pK_b is greater than 14.00 at 15°C.

5. **Indicate which of the following drugs are salts of a weak acid and strong base:**

a. chlorpromazine hydrochloride

b. sodium salicylate

c. acetylsalicylic acid

d. flucloxacillin sodium

e. chlorpheniramine maleate

6. **The solubility of the weakly acidic drug benzylpenicillin (pK_a = 2.76) at pH 8.0 and 20°C is 0.174 mol dm^{-3}. Calculate the solubility when the pH is so low that only the undissociated form of the drug is present in solution. Answer in mol dm^{-3} is:**

a. 0.174×10^{-6}

b. 1.00×10^{-6}

c. 3.024×10^{4}

d. 3.024×10^{-4}

e. 1.00×10^{6}

7. **What is the pH of a solution of ascorbic acid (pK_a = 4.17) of concentration 0.284 mol dm^{-3}?**

a. 1.82

b. 2.64

c. 2.36

d. 1.54

e. 9.36

8. **What is the percentage of promethazine (pK_a = 9.1) existing as free base (i.e. unionised) in a solution of promethazine hydrochloride at pH 7.4?**

a. 98.04

b. 1.96

c. 0.32

d. 99.68

9. **What is the pH of a buffer solution containing 0.025 mol of ethanoic acid (pK_a = 4.76) and 0.035 mol of sodium ethanoate in 1 litre of water?**

a. 6.14

b. 4.59

c. 4.89

d. 2.37

10. Indicate which of the following statements are correct:

a. When a solution is contained within a semipermeable membrane and separated from the solvent, the solvent will pass across the membrane into the solution.

b. The chemical potential of the solvent molecule in the solution is greater than that in pure solvent.

c. A solution that has a greater osmotic pressure than blood is said to be hypertonic.

d. When red blood cells are immersed in a hypotonic solution they will shrink.

11. How many grams of sodium chloride should be added to 25 ml of a 1% solution of tetracaine hydrochloride to make the solution isotonic? The freezing-point depression of 1% tetracaine hydrochloride is 0.109°C, the freezing-point depression of 1% NaCl is 0.576°C and the freezing point of blood serum is –0.52°C.

a. 0.884

b. 0.273

c. 1.092

d. 0.221

12. Indicate which of the following statements are correct:

a. The partition coefficient P is usually defined as the ratio of solubility in the aqueous phase to that in the non-aqueous phase.

b. The greater the value of log P, the higher the lipid solubility of the solute.

c. Ionised solutes will readily partition into the non-aqueous phase.

d. The partition coefficient of an amphoteric drug is at a maximum value at the isoelectric point.

13. Indicate which of the following statements are correct:

a. Drug molecules in solution will diffuse from a region of high chemical potential to one of low chemical potential.

b. The units of diffusion coefficient are m s^{-1}.

c. The diffusion coefficient decreases as the radius of the diffusing molecule increases.

d. The diffusion coefficient decreases when the viscosity of the solution is decreased.

e. The diffusion coefficient increases when the temperature is increased.

chapter 3

Drug stability

Overview

In this chapter we will:
- identify those classes of drugs that are particularly susceptible to chemical breakdown and examine some of the precautions that can be taken to minimise the loss of activity
- look at how reactions can be classified into various orders, and how we can calculate the rate constant for a reaction under a given set of environmental conditions
- look at some of the factors that influence drug stability
- examine methods for accelerating drug breakdown using elevated temperatures and see how to estimate drug stability at the required storage conditions from these measurements.

The chemical breakdown of drugs

The main ways in which drugs break down are as follows:

Hydrolysis
- Drugs containing ester, amide, lactam, imide or carbamate groups are susceptible to hydrolysis.
- Hydrolysis can be catalysed by hydrogen ions (specific acid catalysis) or hydroxyl ions (specific base catalysis).
- Solutions can be stabilised by formulating at the pH of maximum stability or, in some cases, by altering the dielectric constant by the addition of non-aqueous solvents.

Oxidation
- Oxidation involves the removal of an electropositive atom, radical or electron, or the addition of an electronegative atom or radical.
- Oxidative degradation can occur by auto-oxidation, in which reaction is uncatalysed and proceeds quite slowly under the influence of molecular oxygen, or may involve chain processes consisting of three

KeyPoints

- Drugs may break down in solution and also in the solid state (for example, in tablet or powder form).
- It is often possible to predict which drugs are likely to decompose by looking for specific chemical groups in their structures.
- The most common causes of decomposition are hydrolysis and oxidation, but loss of therapeutic activity can also result from isomerisation, photochemical decomposition and polymerisation of drugs.
- It is possible to minimise breakdown by optimising the formulation and storing under carefully controlled conditions.

concurrent reactions: initiation, propagation and termination.

- Examples of drugs that are susceptible to oxidation include steroids and sterols, polyunsaturated fatty acids, phenothiazines, and drugs such as simvastatin and polyene antibiotics that contain conjugated double bonds.
- Various precautions should be taken during manufacture and storage to minimise oxidation:
  - The oxygen in pharmaceutical containers should be replaced with nitrogen or carbon dioxide.
  - Contact of the drug with heavy-metal ions such as iron, cobalt or nickel, which catalyse oxidation, should be avoided.
  - Storage should be at reduced temperatures.
  - Antioxidants should be included in the formulation.

Isomerisation

- Isomerisation is the process of conversion of a drug into its optical or geometric isomers, which are often of lower therapeutic activity.
- Examples of drugs that undergo isomerisation include adrenaline (epinephrine: racemisation in acidic solution), tetracyclines (epimerisation in acid solution), cephalosporins (base-catalysed isomerisation) and vitamin A (*cis–trans* isomerisation).

KeyPoints

- Reactions may be classified according to the order of reaction, which is the number of reacting species whose concentration determines the rate at which the reaction occurs.
- The most important orders of reaction are zero-order (breakdown rate is independent of the concentration of any of the reactants), first-order (reaction rate is determined by one concentration term) and second-order (rate is determined by the concentrations of two reacting species).
- The decomposition of many drugs can occur simultaneously by two or more pathways, which complicates the determination of rate constants.

Photochemical decomposition

- Examples of drugs that degrade when exposed to light include phenothiazines, hydrocortisone, prednisolone, riboflavin, ascorbic acid and folic acid.
- Photodecomposition may occur not only during storage, but also during use of the product. For example, sunlight is able to penetrate the skin to a depth sufficient to cause photodegradation of drugs circulating in the surface capillaries or in the eyes of patients receiving the drug.
- Pharmaceutical products can be adequately protected from photo-induced decomposition by the use of coloured-glass containers (amber glass excludes light of wavelength < 470 nm) and storage in the dark. Coating tablets with a polymer film containing ultraviolet absorbers has been suggested as an additional method for protection from light.

Polymerisation

- Polymerisation is the process by which two or more identical drug molecules combine together to form a complex molecule.
- Examples of drugs that polymerise include amino-penicillins, such as ampicillin sodium in aqueous solution, and also formaldehyde.

Kinetics of chemical decomposition in solution

Zero-order reactions:

- The decomposition proceeds at a constant rate and is independent of the concentrations of any of the reactants.
- The rate equation is:
$$dx/dt = k_0$$
- Integration of the rate equation gives:
$$x = k_0 t$$
- A plot of the amount decomposed (as ordinate) against time (as abscissa) is linear with a slope of k_0 (Figure 3.1).
- The units of k_0 are concentration time^{-1}.
- Many decomposition reactions in the solid phase or in suspensions apparently follow zero-order kinetics.

First-order reactions:

- The rate depends on the concentration of one reactant.
- The rate equation is:
$$dx/dt = k_1(a - x)$$

Tips

- The rate of decomposition of a drug A is the change of concentration of A over a time interval, t, i.e., $-d[A]/dt$ (note that this is negative because the drug concentration is decreasing). However, it is more usual to express the rate as dx/dt, where x is amount of drug which has reacted in time t.
- We can show that $-d[A]/dt$ is equal to dx/dt as follows. If the initial concentration of drug A is a mol dm^{-3} and if we find experimentally that x mol dm^{-3} of the drug has reacted in time t, then the amount of drug remaining at a time t, i.e. $[A]$, is $(a - x)$ mol dm^{-3} and the rate of reaction is:
$$-d[A]/dt = -d(a - x)/dt = dx/dt$$
Notice that the term a is a constant and therefore disappears during differentiation.
- So, when we say that a drug A decomposes by a first-order reaction it means that the rate is proportional to the concentration of A at any particular time, i.e., rate $\propto [A]$. The rate of reaction is therefore given by $dx/dt = k[A]$, where the proportionality constant, k, is called the *rate constant*.

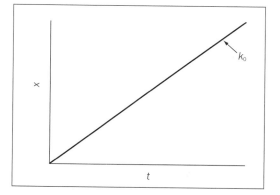

Figure 3.1 Plot of the amount decomposed against time for a zero-order reaction.

- Integration of the rate equation gives:

$$k_1 = \frac{2.303}{t} \log \frac{a}{a-x}$$

- Rearrangement into a linear equation gives:

$$t = \frac{2.303}{k_1} \log a - \frac{2.303}{k_1} \log (a-x)$$

- A plot of time (as ordinate) against the logarithm of the amount remaining (as abscissa) is linear with a slope $= -2.303/k_1$ (Figure 3.2).
- The units of k_1 are time^{-1}.
- If there are two reactants and one is in large excess, the reaction may still follow first-order kinetics because the change in concentration of the excess reactant is negligible. This type of reaction is a *pseudo first-order reaction*.
- The half-life of a first-order reaction is $t_{0.5} = 0.693/k_1$. The half-life is therefore independent of the initial concentration of reactants.

Figure 3.2 Plot of log amount of reactant remaining against time for a first-order reaction

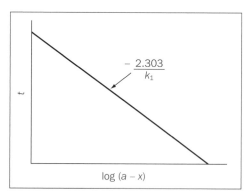

Second-order reactions:

- The rate depends on the concentration of two reacting species, A and B.
- For the usual case where the initial concentrations of A and B are different, the rate equation is:

$$dx/dt = k_2(a-x)(b-x)$$

where a and b are the initial concentrations of reactants A and B, respectively.
- The integrated rate equation is:

$$k_2 = \frac{2.303}{t(a-b)} \log \frac{b(a-x)}{a(b-x)}$$

- Rearrangement into a linear equation gives:

$$t = \frac{2.303}{k_2(a-b)} = \log \frac{b}{a} + \frac{2.303}{k_2(a-b)} \log \frac{(a-x)}{(b-x)}$$

- A plot of time (as ordinate) against the logarithm of $[(a - x)/(b - x)]$ (as abscissa) is linear with a slope $= 2.303/k_2(a - b)$ (Figure 3.3).
- The units of k_2 are concentration^{-1} time^{-1}.
- The half-life of a second-order reaction depends on the initial concentration of reactants and it is not possible to derive a simple expression to calculate it.

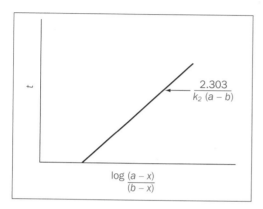

Figure 3.3 Plot of log (amount of reactant A remaining/amount of reactant B remaining) against time for a second-order reaction.

Complex reactions

These are reactions involving simultaneous breakdown by more than one route or by a sequence of reaction steps. Some examples include:

- *Reversible* reactions of the type

$$A \underset{k_r}{\overset{k_f}{\rightleftharpoons}} B$$

where k_f is the rate of the forward reaction and k_r is the rate of the reverse reaction. For these reactions the rate constants can be calculated from a plot of t (as ordinate) against $\log[(A_o - A_{eq})/(A - A_{eq})]$, where A_o, A and A_{eq} represent the initial concentration, the concentration at time t and the equilibrium concentration of reactant A, respectively. The plot should be linear with a slope of $2.303/(k_f + k_r)$. k_f and k_r may be calculated separately if the equilibrium constant K is also determined, since $K = k_f/k_r$.
- *Parallel* reactions in which the decomposition involves two or more pathways, the preference for each route depending on the conditions. Values of the rate constants k_A and k_B for each route may be evaluated separately by determining experimentally the overall rate constant, k_{exp}, and also the ratio R of the concentration of products formed by each reaction from $R = [A]/[B] = k_A/k_B$. It is then possible to calculate the rate constants from $k_A = k_{exp}(R/(R + 1))$ and $k_B = k_A/R$.

Tip

The half-life is the time taken for half the reactant to decompose. Therefore, to derive an expression for the half-life, substitute $t = t_{0.5}$, x $= a/2$, and $(a - x) = a - a/2 = a/2$ into the integrated-rate equation. This now becomes

$$t_{0.5} = \frac{2.303}{k_1} \log \frac{a}{a/2}$$

and hence

$$t_{0.5} = \frac{2.303}{k_1} \log 2 = \frac{0.693}{k_1}$$

Tip

Note that the rate equations used for plotting experimental data are linear equations of the form $y = mx + c$. It is important to remember this when plotting the data. For example, if you are fitting data to the equation

$$t = \frac{2.303}{k_1} \log a - \frac{2.303}{k_1} \log (a-x)$$

then $t = y$, $\log (a-x) = x$ and the gradient $m = -2.303/k_1$. If, for convenience, you plot t on the x-axis and $\log (a-x)$ on the y-axis, then the gradient will, of course, be $-k_1/2.303$.

KeyPoint

The rate of decomposition is influenced by formulation factors such as the pH of the liquid preparation or the addition of electrolytes to control tonicity, and also by environmental factors such as temperature, light and oxygen. An understanding of the way in which these affect the rate of reaction often suggests a means of stabilising the product.

- *Consecutive* reactions in which drug A decomposes to an intermediate B which then decomposes to product C. Each of the decomposition steps has its own rate constant but there is no simple equation to calculate them.

Factors influencing drug stability of liquid dosage forms

pH

- pH has a significant influence on the rate of decomposition of drugs that are hydrolysed in solution and it usual to minimise this effect by formulating at the pH of maximum stability using buffers.
- The rate of reaction is, however, influenced not only by the catalytic effect of hydrogen and hydroxyl ions (*specific acid–base catalysis*), but also by the components of the buffer system (*general acid–base catalysis*). The effect of the buffer components can be large. For example, the hydrolysis rate of codeine in 0.05 M phosphate buffer at pH 7 is almost 20 times faster than in unbuffered solution at this pH.
- The general equation for these two effects is:

$$k_{obs} = k_0 + k_{H^+} [H^+] + k_{OH^-} [OH^-] + k_{HX} [HX] + k_{X^-} [X^-]$$

where k_{obs} is the experimentally determined hydrolytic rate constant, k_0 is the uncatalysed or solvent-catalysed rate constant, k_{H^+} and k_{OH^-} are the specific acid and base catalysis rate constants respectively, k_{HX} and k_{X^-} are the general acid and base catalysis rate constants respectively and [HX] and [X$^-$] denote the concentrations of protonated and unprotonated forms of the buffer.

- The ability of a buffer component to catalyse hydrolysis is related to its dissociation constant, K, by the Brønsted catalysis law. The catalytic coefficient of a buffer component which is a weak acid is given by $k_A = aK_A^\alpha$; the catalytic coefficient of a weak base $k_B = bK_B^\beta$, where a, b, α, and β are constants, and α and β are positive and vary between 0 and 1.

- To remove the influence of the buffer, the reaction rate should be measured at a series of buffer concentrations at each pH and the data extrapolated back to zero buffer concentration. These extrapolated rate constants are plotted as a function of pH to give the required buffer-independent pH–rate profile (Figure 3.4)

Figure 3.4 A typical plot of log rate constant as a function of pH for a drug (codeine sulfate) which undergoes both acid and base catalysis. Modified from M.F. Powell, *J.Pharm. Sci.*75, 901 with permission.

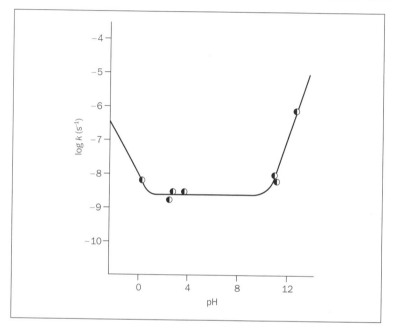

- The rate constants for specific acid and base catalysis can be determined from the linear plots obtained when the corrected experimental rate constants k_{obs} are plotted against the hydrogen ion concentration [H⁺] at low pH (gradient is k_{H^+}), and against the hydroxyl ion concentration at high pH (gradient is k_{OH^-}).
- Complex pH rate profiles are seen when the ionisation of the drug changes over the pH of measurement because of the differing susceptibility of the unionised and ionised forms of the drug to hydrolysis.
- The *oxidative degradation* of some drugs, for example, prednisolone and morphine, in solution may be pH-dependent because of the effect of pH on the oxidation-reduction potential, E_0, of the drug.
- The *photodegradation* of several drugs, for example midazolam and ciprofloxacin, is also pH-dependent.

Temperature

- Increase in temperature usually causes a very pronounced increase in the hydrolysis rate of drugs in solution. This effect is used as the basis for drug stability testing.
- The equation which describes the effect of temperature on decomposition is the *Arrhenius equation*:

$$\log k = \log A - E_a/(2.303RT)$$

where E_a is the activation energy, A is the frequency factor, R is the gas constant (8.314 J mol^{-1} K^{-1}) and T is the temperature in kelvins.
- The Arrhenius equation predicts that a plot of the log rate constant, k, against the reciprocal of the temperature should be linear with a gradient of $-E_a/2.303R$. Therefore, assuming that there is not a change in the order of reaction with temperature, we can measure rates of reaction at high temperatures (where the reaction occurs relatively rapidly) and extrapolate the Arrhenius plots to estimate the rate constant at room temperature (where reaction occurs at a very slow rate) (Figure 3.5).This method therefore provides a means of speeding up the measurements of drug stability during preformulation.

Figure 3.5 A typical Arrhenius plot showing the determination of a rate constant at room temperature by extrapolation of data at high temperatures.

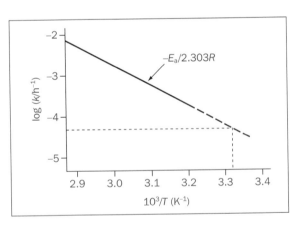

- If a drug formulation is particularly unstable at room temperature, for example, injections of penicillin, insulin, oxytocin and vasopressin, it should be labelled with instructions to store in a cool place.

Ionic strength

- The equation which describes the influence of electrolyte on the rate constant is the *Brønsted–Bjerrum equation*:

$$\log k = \log k_0 + 2Az_A z_B \sqrt{\mu}$$

where z_A and z_B are the charge numbers of the two interacting ions, A is a constant for a given solvent and temperature and μ is the ionic strength.

- The Brønsted–Bjerrum equation predicts that a plot of $\log k$ against $\mu^{1/2}$ should be linear for a reaction in the presence of different concentrations of the same electrolyte with a gradient of $2Az_A z_B$ (Figure 3.6).
- The gradient will be positive (i.e. the reaction rate will be increased by electrolyte addition) when reaction is between ions of similar charge, for example, the acid-catalysed hydrolysis of a cationic drug ion.
- The gradient will be negative (i.e. the reaction rate will be decreased by electrolyte addition) when the reaction is between ions of opposite charge, for example, the base-catalysed hydrolysis of positively charged drug species.

Tip

Ionic strength can be calculated from:
$$\mu = \tfrac{1}{2} \Sigma(mz^2)$$
$$= \tfrac{1}{2}(m_A z_A^2 + m_B z_B^2 + \ldots)$$
So, for example, if we have a monovalent drug ion of concentration 0.01 mol kg^{-1} in the presence of 0.001 mol kg^{-1} of Ca^{2+} ions, then the ionic strength of the solution will be $\mu = \tfrac{1}{2}[(0.01 \times 1^2) + (0.001 \times 2^2)] = 0.007$ mol kg^{-1}. Note that if the drug ion and the electrolyte ion are both monovalent, then the ionic strength will be equal to the total molality of the solution.

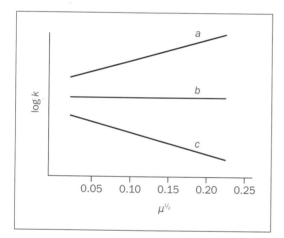

Figure 3.6 The variation of rate constant, k, with square root of ionic strength, μ, for reaction between a: ions of similar charge, b: ion and uncharged molecule and c: ions of opposite charge.

Solvent effects

- The equation that describes the effect of the dielectric constant, ε, on the rate of hydrolysis is:
$$\log k = \log k_{\varepsilon=\infty} - Kz_A z_B / \varepsilon$$
where K is a constant for a particular reaction at a given temperature, z_A and z_B are the charge numbers of the two interacting ions and $k_{\varepsilon=\infty}$ is the rate constant in a theoretical solvent of infinite dielectric constant.
- This equation predicts that a plot of $\log k$ against the reciprocal of the dielectric constant of the solvent should be linear with a gradient $-Kz_A z_B$. The intercept when $1/\varepsilon = 0$ (i.e. when $\varepsilon = \infty$) is equal to the logarithm of the rate constant, $k_{\varepsilon=\infty}$, in a theoretical solvent of infinite dielectric constant (Figure 3.7):

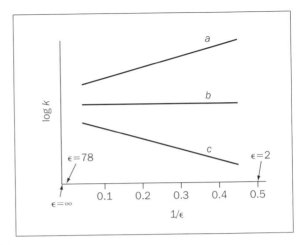

Figure 3.7 The variation of rate constant with reciprocal of dielectric constant for reaction between *a*, ions of opposite charge, *b*, ion and uncharged molecule and *c*, ions of similar charge.

Tips

- The dielectric constant (or relative permittivity) of a solvent is a measure of its polarity. Water has a high dielectric constant (approximately 78 at room temperature); other solvents have much lower values (for example, ε of ethanol is approximately 24).
- Obviously, the particular solvent chosen to replace water in non-aqueous formulations must be non-toxic and alcohol–water or propylene glycol–water mixtures may be suitable for this purpose.

- The gradient will be negative when the charges on the drug ion and the interacting species are the same. This means that if we replace the water with a solvent of lower dielectric constant then we will achieve the desired effect of reducing the reaction rate.
- The gradient will be positive if the drug ion and the interacting ion are of opposite signs and therefore the choice of a non-polar solvent will only result in an increase of decomposition.

Oxygen

- The susceptibility of a drug to the presence of oxygen can be tested by comparing its stability in ampoules purged with oxygen to that when it is stored under nitrogen.
- Drugs which have a higher rate of decomposition when exposed to oxygen can be stabilised by replacing the oxygen in the storage container with nitrogen or carbon dioxide. These drugs should also be kept out of contact with heavy metals and should be stabilised with antioxidants.

Light

- The susceptibility of a drug to light can readily be tested by comparing its stability when exposed to light to that when stored in the dark.
- Photolabile drugs should be stored in containers of amber glass and, as an added precaution, should be kept in the dark.

Factors influencing drug stability of solid dosage forms

Moisture

- Water-soluble drugs in the solid dosage form will dissolve in any moisture layer which forms on the solid surface. The drug will now be in an aqueous environment and will be affected by many of the same factors as for liquid dosage forms.
- It is important to select packaging that will exclude moisture during storage.

Excipients

- Excipients such as starch and povidone have high water contents and affect stability by increasing the water content of the formulation.
- Chemical interactions between the excipients and the drug can sometimes occur and these lead to a decrease of stability. For example, stearate salts used as tablet lubricants can cause base-catalysed hydrolysis; polyoxyethlene glycols used as suppository bases can cause degradation of aspirin.

Temperature

- The effect of temperature on stability can sometimes be described by the Arrhenius equation, but complications arise if the dosage form melts on temperature increase (e.g. suppositories) or if the drug or one of the excipients changes its polymorphic form.

Light and oxygen

- Solid dosage formulations containing photolabile drugs or drugs susceptible to oxidation should be stored in the same way as described for liquid dosage forms to protect from light and oxygen. Note also that moisture contains dissolved oxygen and hence the preparations should be stored in dry conditions.

KeyPoints

- It is most important to be able to ensure that a particular formulation when packaged in a specific container will remain within its physical, chemical, microbiological, therapeutic and toxicological specifications on storage for a specified time period.
- In order to have such an assurance we need to conduct a rigorous stability testing programme on the product in the form that is finally to be marketed.
- To calculate the shelf-life it is necessary to know the rate constant at the storage temperature. However, the rate of breakdown of most pharmaceutical products is so slow that it would take many months to determine this at room temperature and it has become essential to devise a more rapid technique which can be used during product development to speed up the identification of the most suitable formulation.
- The method that is used for accelerated storage testing is based on the Arrhenius equation.

Stability testing and calculation of shelf-life

The Arrhenius equation is used as the basis of a method for accelerating decomposition by raising the temperature of the preparations. This method provides a means for rapidly identifying the most suitable preparation during preformulation of the product. The main steps in the process are:

- Determination of the order of reaction by plotting stability data at several elevated temperatures according to the equations relating decomposition to time for each of the orders of reaction, until linear plots are obtained.
- Values of the rate constant k at each temperature are calculated from the gradient of these plots, and the logarithm of k is plotted against reciprocal temperature according to the Arrhenius equation $\log k = \log A - E_a/2.303RT$.
- A value of k can be interpolated from this plot at the required temperature.
- Alternatively, if only an approximate value of k is required at temperature T_1, then this may be estimated from measurements at a single higher temperature T_2 using

$$\log\left[\frac{k_2}{k_1}\right] = \frac{E_a(T_2 - T_1)}{2.303RT_2T_1}$$

where k_1 and k_2 are the rate constants at temperatures T_1 and T_2 respectively. A mid-range value of $E_a = 75$ kJ mol^{-1} may be used for these rough estimations.

- The shelf-life for the product can be calculated from the rate constant based on an acceptable degree of decomposition. For example, for decomposition which follows first-order kinetics, the time taken for 10% loss of activity is given by $t_{90} = 0.105/k_1$.

> **Tip**
>
> You can derive an expression for the time taken for 10% of the reactant to decompose by substituting $t = t_{0.9}$, $x = 0.1a$, $(a - x) = a - 0.1a = 0.9a$ into the integrated-rate equation for the relevant order of reaction. For example, substituting into the first-order rate equation gives
>
> $$t_{0.9} = \frac{2.303}{k_1}\log\frac{a}{0.9a} = \frac{0.105}{k_1}$$

Multiple choice questions

In questions 1–6 indicate whether each of the statements is *true* or *false*.

1. **In a zero-order reaction:**
   a. The rate of decomposition is independent of the concentration of the reactants.
   b. The rate of decomposition is dependent on the concentration of one of the reactants.
   c. A plot of the amount remaining (as ordinate) against time (as abscissa) is linear with a slope of $1/k$.
   d. The units of k are (concentration^{-1} time^{-1}).
   e. The half-life is $t_{0.5} = 0.693/k$.

2. **In a second-order reaction:**
a.   The rate of reaction depends on the concentration of two reacting species.
b.   A plot of time (as ordinate) against the logarithm of $[(a - x)/(b - x)]$ (as abscissa) is linear with a slope $= 2.303/k_2(a - b)$.
c.   The units of k are concentration $\times$ time^{-1}.
d.   The half-life is independent of the concentration of the reactants.

3   **In a study of the hydrolysis of a drug in aqueous solution a plot of logarithm of the amount of drug remaining (as ordinate) against time (as abscissa) is linear.**
a.   The reaction is zero-order.
b.   The slope is $-2.303/k$.
c.   The units of k are (concentration^{-1} time^{-1}).
d.   The half-life is $t_{0.5} = 0.693/k$.
e.   The half-life depends on the initial concentration of reactant.

4. **The Arrhenius equation for effect of temperature on the hydrolysis of a drug in aqueous solution:**
a.   predicts that the rate of reaction will decrease as temperature is increased
b.   predicts that a plot of log k against temperature will be linear
c.   predicts that a plot of log k against the reciprocal of temperature will be linear
d.   predicts that there will be no change in the order of reaction when temperature is increased
e.   is the basis of drug stability testing.

5   **The Brønsted–Bjerrum equation:**
a.   describes the influence of electrolyte on the rate constant
b.   predicts that a plot of log k against ionic strength will be linear
c.   predicts that a plot of log k against the reciprocal of ionic strength will be linear
d.   predicts that the reaction rate will be increased by electrolyte addition when reaction is between ions of similar charge
e.   predicts that the reaction rate will be decreased by electrolyte addition when the reaction is between ions of opposite charge.

6. **The equation that describes the effect of dielectric constant on rate of reaction predicts that:**
a.   A plot of log k against the reciprocal of the dielectric constant of the solvent should be linear.
b.   A plot of k against the reciprocal of the dielectric constant of the solvent should be linear.
c.   Replacing the water with a solvent of lower dielectric constant will always reduce the reaction rate.
d.   The rate of hydrolysis will increase when a less polar solvent is used if the drug ion and the interacting ion are of opposite charge.
e.   The rate of hydrolysis will increase when a less polar solvent is used if the charges on the drug ion and the interacting species are the same.

7. **Indicate which of the following statements relating to the effect of pH on drug stability are true:**
a.   The rate of acid-catalysed decomposition of a drug increases with pH.
b.   The rate of base-catalysed decomposition of a drug increases with the concentration of hydroxyl ions.
c.   The effect of buffer components on decomposition is referred to as general acid–base catalysis.
d.   A plot of the observed-rate constant (as ordinate) against pH (as abscissa) for an acid-catalysed reaction has a gradient equal to k_H^+.

8. **What is the remaining concentration $(a - x)$ in mg ml^{-1} of a drug (initial concentration $a = 7$ mg ml^{-1}) after a time equivalent to 3 half-lives assuming that the decomposition follows first-order kinetics?**
a.   2.33
b.   3.5
c.   1.75
d.   0.875
e.   1.167

9. **The time taken for 5% of a drug to decompose by first-order kinetics is:**
a. $0.022/k_1$
b. $0.051/k_1$
c. $0.105/k_1$
d. $k_1/0.051$
e. $0.105\ k_1$

chapter 4
Surfactants

Overview

In this chapter we will:

- see why certain molecules have the ability to lower the surface and interfacial tension and how the surface activity of a molecule is related to its molecular structure
- look at the properties of some surfactants that are commonly used in pharmacy
- examine the nature and properties of monolayers formed when insoluble surfactants are spread over the surface of a liquid
- look at some of the factors that influence adsorption onto solid surfaces and see how experimental data from adsorption experiments may be analysed to gain information on the process of adsorption
- see why micelles are formed, examine the structure of ionic and non-ionic micelles and look at some of the factors that influence micelle formation
- examine the properties of liquid crystals and surfactant vesicles
- discuss the process of solubilisation of water-insoluble compounds by surfactant micelles and its applications in pharmacy.

Some typical surfactants

Depending on their charge characteristics the surface-active molecules may be anionic, cationic, zwitterionic (ampholytic) or non-ionic. Examples of surfactants that are used in pharmaceutical formulation are as follows:

Anionic surfactants: Sodium Lauryl Sulphate BP

- is a mixture of sodium alkyl sulfates, the chief of which is sodium dodecyl sulfate, $C_{12}H_{25}SO_4^- Na^+$
- is very soluble in water at room temperature, and is used pharmaceutically as a preoperative skin cleaner, having bacteriostatic action against gram-positive bacteria, and also in medicated shampoos
- is a component of emulsifying wax.

Cationic surfactants

- The quaternary ammonium and pyridinium cationic surfactants are important pharmaceutically because of their

> **KeyPoint**
>
> Surfactants have two distinct regions in their chemical structure, one of which is water-liking or *hydrophilic* and the other of which is water-hating or *hydrophobic*. These molecules are referred to as *amphiphilic* or *amphipathic* molecules or simply as *surfactants* or *surface active agents*.

bactericidal activity against a wide range of gram-positive and some gram-negative organisms.

- They may be used on the skin, especially in the cleaning of wounds.
- Their aqueous solutions are used for cleaning contaminated utensils.

Non-ionic surfactants

- *Sorbitan esters* are supplied commercially as *Spans* and are mixtures of the partial esters of sorbitol and its mono- and di-anhydrides with oleic acid. They are generally insoluble in water (low hydrophile–lipophile balance (HLB) value) and are used as water-in-oil emulsifiers and as wetting agents.

> **Tip**
>
> HLB stands for hydrophile–lipophile balance. Compounds with a high HLB (greater than about 12) are predominantly hydrophilic and water-soluble. Those with very low HLB values are hydrophobic and water-insoluble.

- *Polysorbates* are complex mixtures of partial esters of sorbitol and its mono- and di-anhydrides condensed with an approximate number of moles of ethylene oxide. They are supplied commercially as *Tweens*. The polysorbates are miscible with water, as reflected in their higher HLB values, and are used as emulsifying agents for oil-in-water emulsions.
- *Poloxamers* are synthetic block copolymers of hydrophilic poly(oxyethylene) and hydrophobic poly(oxypropylene) with the general formula $E_m P_n E_m$, where E = oxyethylene (OCH_2CH_2) and P = oxypropylene ($OCH_2CH(CH_3)$) and the subscripts m and n denote chain lengths. Properties such as viscosity, HLB and physical state (liquid, paste or solid) are dependent on the relative chain lengths of the hydrophilic and hydrophobic blocks. They are supplied commercially as *Pluronics* and are labelled using the Pluronic grid, for example as F127 or L62, where the letter indicates the physical state (F, P or L, denoting solid, paste or liquid, respectively). The last digit of this number is approximately one-tenth of the weight percentage of poly(oxyethylene); the first one (or two digits in a three-digit number) multiplied by 300 gives a rough estimate of the molecular weight of the hydrophobe.

A wide variety of drugs, including the antihistamines and the tricyclic depressants, are surface-active.

Reduction of surface and interfacial tension

When surfactants are dissolved in water they orientate at the surface so that the hydrophobic regions are removed from the aqueous environment, as shown in Figure 4.1a. The reason for the reduction in the surface tension when surfactant molecules adsorb at the water surface is that the surfactant molecules replace some of the water molecules in the surface and the forces of attraction between surfactant and water molecules are less than those between two water molecules, hence the contraction force is reduced.

Figure 4.1 Orientation of amphiphiles at (a) solution–vapour interface and (b) hydrocarbon–solution interface.

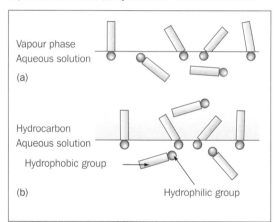

Surfactants will also adsorb at the interface between two immiscible liquids such as oil and water and will orientate themselves as shown in Figure 4.1b, with their hydrophilic group in the water and their hydrophobic group in the oil. The interfacial tension at this interface, which arises because of a similar imbalance of attractive forces as at the water surface, will be reduced by this adsorption.

There is an equilibrium between surfactant molecules at the surface of the solution and those in the bulk of the solution which is expressed by the Gibbs equation:

$$\Gamma_2 = -\frac{1}{xRT}\frac{d\gamma}{2.303\,d\log c}$$

where Γ_2 is the surface excess concentration, R is the gas constant

Tip

When substituting values into equations it is important to convert the values into the correct units. In the case of the Gibbs equation it is easy to forget to convert concentration into mol m⁻³ (1 mol l⁻¹ = 1 mol dm⁻³ = 10³ mol m⁻³).

(8.314 J mol^{-1} K^{-1}), T is temperature in kelvins, c is the concentration in mol m^{-3} and x has a value of 1 for ionic surfactants in dilute solution.

The area A occupied by a surfactant molecule at the solution–air interface can be calculated from $A = 1/N_A \Gamma_2$ where N_A is the Avogadro number (6.023×10^{23} molecules mol^{-1}) and $d\gamma/d\log c$ is the gradient of the plot of surface tension against $\log c$ measured at a concentration just below the critical micelle concentration (CMC).

The surface activity of a particular surfactant depends on the balance between its hydrophilic and hydrophobic properties. For a homologous series of surfactants:

- An increase in the length of the hydrocarbon chain (hydrophobic) increases the surface activity. This relationship between hydrocarbon chain length and surface activity is expressed by *Traube's rule*, which states that 'in dilute aqueous solutions of surfactants belonging to any one homologous series, the molar concentrations required to produce equal lowering of the surface tension of water decreases threefold for each additional CH$_2$ group in the hydrocarbon chain of the solute'.
- An increase of the length of the ethylene oxide chain (hydrophilic) of a polyoxyethylated non-ionic surfactant results in a decrease of surface activity.

Insoluble monolayers

Insoluble amphiphilic compounds, for example surfactants with very long hydrocarbon chains, will also form films on water surfaces when the amphiphilic compound is dissolved in a volatile solvent and carefully injected onto the surface. Polymers and proteins may also form insoluble monolayers.

The molecules are orientated at the surface in the same way as typical surfactants, i.e. with the hydrophobic group protruding into the air and the polar group acting as an anchor in the surface.

The properties of the film can be studied using a Langmuir trough (Figure 4.2) and the results are presented as plots of surface pressure π ($\pi = \gamma_o - \gamma_m$, where γ_o is the surface tension of the clean surface and γ_m is the surface tension of the film-covered surface) against area per molecule.

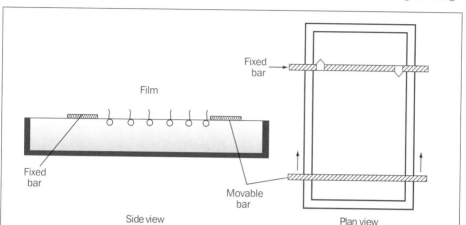

Figure 4.2 Langmuir trough.

There are three main types of insoluble monolayers (Figure 4.3):

- *Solid or condensed monolayers*, in which the film pressure remains very low at high film areas and rises abruptly when the molecules become tightly packed on compression. The extrapolated limiting surface area is very close to the cross-sectional area of the molecule from molecular models.
- *Expanded monolayers*, in which the π–A plots are quite steeply curved but extrapolation to a limiting surface area yields a value that is usually several times greater than the cross-sectional area from molecular models. Films of this type tend to be formed by molecules in which close packing into condensed films is prohibited by bulky side chains or by a *cis* configuration of the molecule.
- *Gaseous monolayers*, in which there is only a gradual change in the surface pressure as the film is compressed. The molecules in this type of monolayer lie along the surface, often because they possess polar groups that are distributed about the molecule and anchor the molecules to the surface along its length. Monolayers of polymers and proteins are often of this type.

Monolayers are useful models by which the properties of polymers used as packaging materials can be investigated. They may also be used as cell membrane models.

Figure 4.3 Surface pressure, π, versus area per molecule, A, for the three main types of monolayer.

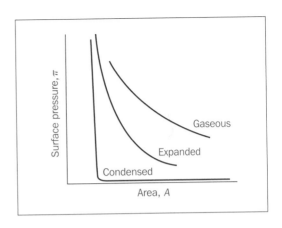

KeyPoint

Note the difference between the term *adsorption*, which is used to describe the process of accumulation at an interface, and *absorption*, which means the penetration of one component throughout the body of a second.

Adsorption at the solid–liquid interface

There are two general types of adsorption:
1. Physical adsorption, in which the adsorbate is bound to the surface through the weak van der Waals forces.
2. Chemical adsorption or chemisorption, which involves the stronger valence forces.

Frequently both physical and chemical adsorption may be involved in a particular adsorption process.

A simple experimental method of studying adsorption is to shake a known mass of the adsorbent material with a solution of known concentration at a fixed temperature until no further change in the concentration of the supernatant is observed, that is, until equilibrium conditions have been established.

Adsorption data may be analysed using the Langmuir and Freundlich equations:

■ The *Langmuir* equation is:
$$x/m = abc/(1 + bc)$$
where x is the amount of solute adsorbed by a weight, m, of adsorbent, c is the concentration of solution at equilibrium, b is a constant related to the enthalpy of adsorption and a is related to the surface area of the solid. For practical usage the Langmuir equation is rearranged into a linear form as:
$$c/(x/m) = 1/ab + c/a$$
Values of a and b may then be determined from the intercept ($1/ab$) and slope ($1/a$) of plots of $c/(x/m)$ against concentration (Figure 4.4).

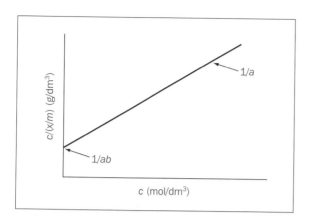

Figure 4.4 A typical Langmuir plot.

- The *Freundlich* equation is:
$$x/m = ac^{1/n}$$
where a and n are constants, the form $1/n$ being used to emphasise that c is raised to a power less than unity. $1/n$ is a dimensionless parameter and is related to the intensity of drug adsorption. The linear form of this equation is:
$$\log (x/m) = \log a + (1/n) \log c$$
A plot of log (x/m) against log c should be linear, with an intercept of log a and slope of $1/n$ (Figure 4.5). It is generally assumed that, for systems that obey this equation, adsorption results in the formation of multilayers rather than a single monolayer.

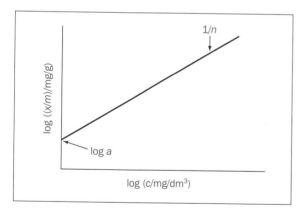

Figure 4.5 A typical Freundlich plot.

Factors affecting adsorption

- *Solubility of the adsorbate.* In general, the extent of adsorption of a solute is inversely proportional to its solubility in the solvent from which adsorption occurs. This empirical rule is termed *Lundelius' rule.* For homologous series, adsorption from solution increases as the series is ascended and the molecules become more hydrophobic.

■ *pH.* In general, for simple molecules adsorption increases as the ionisation of the drug is suppressed, the extent of adsorption reaching a maximum when the drug is completely unionised (Figure 4.6).

■ *Nature of the adsorbent.* The most important property affecting adsorption is the surface area of the adsorbent; the extent of adsorption is proportional to the specific surface area. Thus, the more finely divided or the more porous the solid, the greater will be its adsorption capacity.

■ *Temperature.* Since adsorption is generally an exothermic process, an increase in temperature normally leads to a decrease in the amount adsorbed.

Figure 4.6 The adsorption onto nylon of a typical weakly basic drug (○) and its percentage in unionised form (●) as a function of pH. Reproduced from N.E. Richards and B.J. Meakin. *J. Pharm. Pharmacol.,* 26, 166 (1974) with permission.

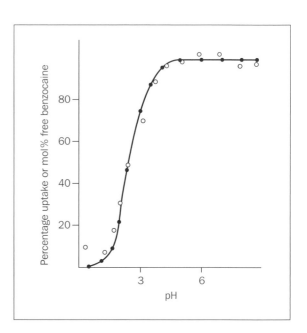

Pharmaceutical applications and consequences of adsorption

■ *Adsorption of poisons/toxins.* The 'universal antidote' for use in reducing the effects of poisoning by the oral route is composed of activated charcoal, magnesium oxide and tannic acid. A more recent use of adsorbents has been in dialysis to reduce toxic concentrations of drugs by passing blood through a haemodialysis membrane over charcoal and other adsorbents.

■ *Taste masking.* Drugs such as diazepam may be adsorbed onto solid substrates to minimise taste problems, but care should be taken to ensure that desorption does not become a rate-limiting step in the absorption process.

- *Haemoperfusion.* Carbon haemoperfusion is an extracorporeal method of treating cases of severe drug overdoses and originally involved perfusion of the blood directly over charcoal granules. Activated charcoal granules are very effective in adsorbing many toxic materials, but they give off embolising particles and also lead to removal of blood platelets. These problems are removed by microencapsulation of activated charcoal granules by coating with biocompatible membranes such as acrylic hydrogels.
- *Adsorption in drug formulation.* Beneficial uses include adsorption of surfactants and polymers in the stabilisation of suspensions, and adsorption of surfactants onto poorly soluble solids to increase their dissolution rate through increased wetting. Problems may arise from the adsorption of medicaments by adsorbents such as antacids, which may be taken simultaneously by the patient, or which may be present in the same formulation; and from the adsorption of medicaments on to the container walls, which may affect the potency and possibly the stability of the product.

Micellisation

Micelles are formed at the *critical micelle concentration* (CMC), which is detected as an inflection point when physicochemical properties such as surface tension are plotted as a function of concentration (Figure 4.7).

The main reason for micelle formation is the attainment of a minimum free energy state. The main driving force for the formation of micelles is the increase of entropy that occurs when the hydrophobic regions of the surfactant are removed from water and the ordered structure of the water molecules around this region of the molecule is lost.

Most micelles are spherical and contain between 60 and 100 surfactant molecules.

KeyPoints

- Micelles are formed at the CMC.
- Micelles are dynamic structures and are continually formed and broken down in solution – they should not be thought of as solid spheres.
- The typical micelle diameter is about 2–3 nm and so they are not visible under the light microscope.
- There is an equilibrium between micelles and free surfactant molecules in solution.
- When the surfactant concentration is increased above the CMC, the number of micelles increases but the free surfactant concentration stays constant at the CMC value.

Tips

- Entropy is a thermodynamic property that is a measure of the randomness or disorder of a system.
- When a system becomes more chaotic its entropy increases, so the loss of water structure when micelles are formed will increase entropy.
- Entropy change ΔS is linked to free energy change ΔG by the equation $\Delta G = \Delta H - T\Delta S$.
- The enthalpy change ΔH when micelles are formed is very small and can be ignored, so you can see that an increase of entropy will lead to a decrease in free energy.
- Any change that leads to a free energy decrease will occur spontaneously because it leads to the formation of a more stable system. Micelle formation is therefore a spontaneous process.

Figure 4.7 Typical plot of the surface tension against logarithm of surfactant concentration, c, showing the critical micelle concentration.

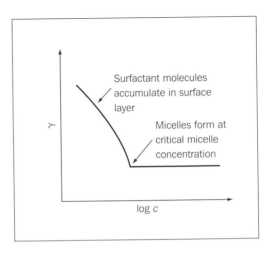

The structure of the micelles formed by *ionic* surfactants (Figure 4.8a) consists of:

- a hydrophobic *core* composed of the hydrocarbon chains of the surfactant molecule
- a *Stern layer* surrounding the core, which is a concentric shell of hydrophilic head groups with $(1 - \alpha)N$ counterions, where α is the degree of ionisation and N is the aggregation number (number of molecules in the micelle). For most ionic micelles the degree of ionisation α is between 0.2 and 0.3; that is, 70–80% of the counterions may be considered to be bound to the micelles
- a *Gouy–Chapman electrical double layer* surrounding the Stern layer, which is a diffuse layer containing the αN counterions required to neutralise the charge on the kinetic micelle. The thickness of the double layer is dependent on the ionic strength of the solution and is greatly compressed in the presence of electrolyte.

Micelles formed by *non-ionic* surfactants:

- are larger than their ionic counterparts and may sometimes be elongated into an ellipsoid or rod-like structure
- have a hydrophobic core formed from the hydrocarbon chains of the surfactant molecules surrounded by a shell (the *palisade layer*) composed of the oxyethylene chains of the surfactant (Figure 4.8b), which is heavily hydrated.

Micelles formed in non-aqueous solution (reverse or inverted micelles) have a core composed of the hydrophilic groups surrounded by a shell of the hydrocarbon chains (Figure 4.8c).

Figure 4.8 (a) Partial cross-section of an anionic micelle showing charged layers; (b) cross-section of a non-ionic micelle; (c) diagrammatic representation of a reverse micelle.

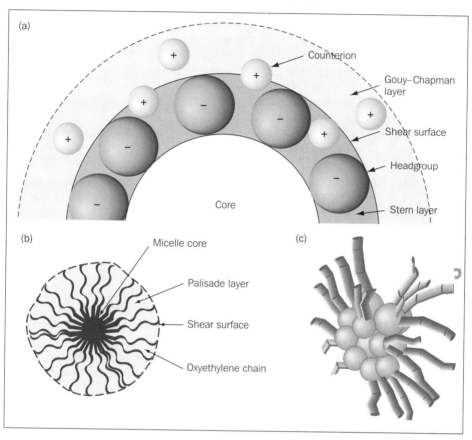

Factors affecting the CMC and micellar size
Structure of the hydrophobic group
Increase in length of the hydrocarbon chain results in:
- a decrease in CMC, which for compounds with identical polar head groups is expressed by the linear equation:

$$\log [\text{CMC}] = A - Bm$$

where m is the number of carbon atoms in the chain and A and B are constants for a homologous series.
- a corresponding increase in micellar size.

Nature of the hydrophilic group
- Non-ionic surfactants generally have very much lower CMC values and higher aggregation numbers than their ionic counterparts with similar hydrocarbon chains.
- An increase in the ethylene oxide chain length of a non-ionic surfactant makes the molecule more hydrophilic and the CMC increases.

Type of counterion

- Micellar size increases for a particular cationic surfactant as the counterion is changed according to the series $Cl^- < Br^- < I^-$, and for a particular anionic surfactant according to $Na^+ < K^+ < Cs^+$.
- Ionic surfactants with organic counterions (e.g. maleates) have lower CMCs and higher aggregation numbers than those with inorganic counterions.

Addition of electrolytes

- Electrolyte addition to solutions of ionic surfactants decreases the CMC and increases the micellar size. This is because the electrolyte reduces the forces of repulsion between the charged head groups at the micelle surface, so allowing the micelle to grow.
- At high electrolyte concentration the micelles of ionic surfactants may become non-spherical.

Effect of temperature

- Aqueous solutions of many non-ionic surfactants become turbid at a characteristic temperature called the *cloud point*.
- At temperatures up to the cloud point there is an increase in micellar size and a corresponding decrease in CMC.
- Temperature has a comparatively small effect on the micellar properties of ionic surfactants.

KeyPoints

- The properties of a surfactant are determined by the balance between the hydrophobic and hydrophilic parts of the molecule.
- If the hydrophobic chain length is increased then the whole molecule becomes more hydrophobic and micelles will form at lower solution concentration, i.e. the CMC decreases.
- If the hydrophilic chain length is increased then the molecule becomes more hydrophilic and the CMC will increase.

Formation of liquid crystals and vesicles

Lyotropic liquid crystals

The liquid crystalline phases that occur on increasing the concentration of surfactant solutions are referred to as *lyotropic* liquid crystals; their structure is shown diagrammatically in Figure 4.9.

Figure 4.9 Diagrammatic representation of forms of lyotropic liquid crystals.

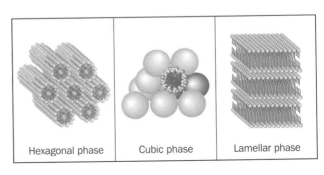

| Hexagonal phase | Cubic phase | Lamellar phase |

- Increase of concentration of a surfactant solution frequently causes a transition from the typical spherical micellar structure to a more elongated or rod-like micelle.
- Further increase in concentration may cause the orientation and close packing of the elongated micelles into hexagonal arrays; this is a liquid crystalline state termed the *middle phase* or *hexagonal phase*.
- With some surfactants, further increase of concentration results in the separation of a second liquid crystalline state – the *neat phase* or *lamellar phase*.
- In some surfactant systems another liquid crystalline state, the *cubic phase*, occurs between the middle and neat phases (Figure 4.10).

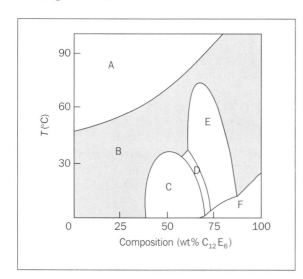

Figure 4.10 Phase diagram of a typical non-ionic surfactant in water. A, two isotropic liquid phases; B, micellar solution; C, middle or hexagonal phase; D, cubic phase; E, neat or lamellar phase; F, solid phase. The boundary between phases A and B is the cloud point. Modified from Clunie J S, Goodman J F, Symons P C. *Trans Farad. Soc.* 1969; 65: 287.

The lyotropic liquid crystals are anisotropic, that is, their physical properties vary with direction of measurement.

- The middle phase, for example, will only flow in a direction parallel to the long axis of the arrays. It is rigid in the other two directions.
- The neat phase is more fluid and behaves as a solid only in the direction perpendicular to that of the layers.
- Plane-polarised light is rotated when travelling along any axis except the long axis in the middle phase and a direction perpendicular to the layers in the neat phase.

KeyPoint

Because of their ability to rotate polarised light, liquid crystals are visible when placed between crossed polarisers and this provides a useful means of detecting the liquid crystalline state.

Thermotropic liquid crystals

Thermotropic liquid crystals are produced when certain substances, for example the esters of cholesterol, are heated. The arrangement of the elongated molecules in thermotropic liquid crystals is generally recognisable as one of three principal types (Figure 4.11):

1. *Nematic liquid crystals*:
   - Groups of molecules orientate spontaneously with their long axes parallel, but they are not ordered into layers.
   - Because the molecules have freedom of rotation about their long axis, the nematic liquid crystals are quite mobile and are readily orientated by electric or magnetic fields.
2. *Smectic liquid crystals*:
   - Groups of molecules are arranged with their long axes parallel, and are also arranged into distinct layers.
   - As a result of their two-dimensional order the smectic liquid crystals are viscous and are not orientated by magnetic fields.
3. *Cholesteric (or chiral nematic) liquid crystals*:
   - Are formed by several cholesteryl esters.
   - Can be visualised as a stack of very thin two-dimensional nematic-like layers in which the elongated molecules lie parallel to each other in the plane of the layer.
   - The orientation of the long axes in each layer is displaced from that in the adjacent layer and this displacement is cumulative through successive layers so that the overall displacement traces out a helical path through the layers.
   - The helical path causes very pronounced rotation of polarised light, which can be as much as 50 rotations per millimeter.
   - The pitch of the helix (the distance required for one complete rotation) is very sensitive to small changes in temperature and pressure and dramatic colour changes can result from variations in these properties.
   - The cholesteric phase has a characteristic iridescent appearance when illuminated by white light due to circular dichroism.

Vesicles

Vesicles are formed by phospholipids and other surfactants having two hydrophobic chains. There are several types:

Liposomes

- Liposomes are formed by naturally occurring phospholipids such as lecithin (phosphatidyl choline).
- They can be multilamellar (composed of several bimolecular lipid lamellae separated by aqueous layers) or unilamellar (formed by sonication of solutions of multilamellar liposomes).

Figure 4.11 Diagrammatic representation of forms of thermotropic liquid crystals. (a) smectic, (b) nematic and (c) cholesteric liquid crystals.

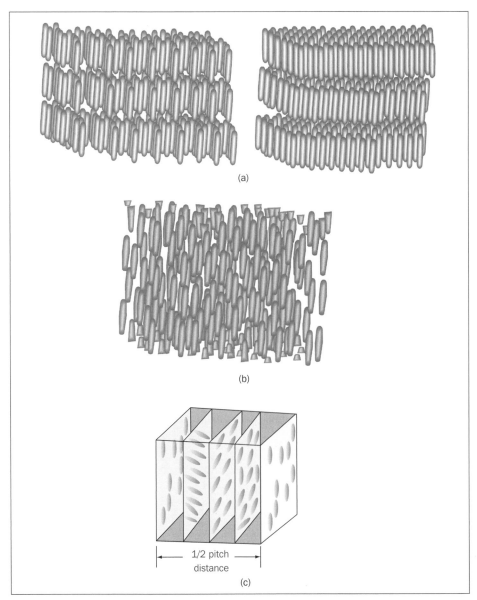

(a)

(b)

1/2 pitch distance

(c)

■ They may be used as drug carriers; water-soluble drugs can be entrapped in liposomes by intercalation in the aqueous layers, whereas lipid-soluble drugs can be solubilised within the hydrocarbon interiors of the lipid bilayers.

Surfactant vesicles and niosomes
■ Formed by surfactants having two alkyl chains.
■ Sonication can produce single-compartment vesicles.

KeyPoints

- Solubilisation is the process whereby water-insoluble substances are brought into solution by incorporation into micelles.
- There is a difference between this exact use of the term 'solubilisation' which is used in this chapter and its more general use to mean simply to dissolve in solution.

- Vesicles formed by ionic surfactants are useful as membrane models.
- Vesicles formed from non-ionic surfactants are called niosomes and have potential use in drug delivery.

Solubilisation

The *maximum amount of solubilisate* that can be incorporated into a given system at a fixed concentration is termed the *maximum additive concentration* (MAC).

Solubility data are expressed as a solubility versus concentration curve or as three-component phase diagrams, which describe the effect of varying all three components of the system (solubilisate, solubiliser and solvent).

The *site of solubilisation* within the micelle is closely related to the chemical nature of the solubilisate (Figure 4.12):

- Non-polar solubilisates (aliphatic hydrocarbons, for example) are dissolved in the hydrocarbon core of ionic and non-ionic micelles (position 1).
- Water-insoluble compounds containing polar groups are orientated with the polar group at the core–surface interface of the micelle, and the hydrophobic group buried inside the hydrocarbon core of the micelle (position 2 and 3).
- In addition to these sites, solubilisation in non-ionic polyoxyethylated surfactants can also occur in the poly–oxyethylene shell (palisade layer) which surrounds the core (position 4).

Figure 4.12 Schematic representation of sites of solubilisation depending on the hydrophobicity of the solubilisate. Redrawn from Torchilin V. *J Control Release* 2001; 73: 137.

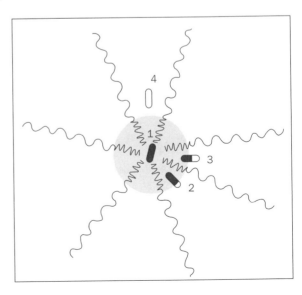

Factors affecting solubilisation capacity

Nature of the surfactant

- When the solubilisate is located within the core or deep within the micelle structure the solubilisation capacity increases with increase in alkyl chain length up to about C_{16}; further increase has little effect on solubilisation capacity.
- The effect of an increase in the ethylene oxide chain length of a polyoxyethylated non-ionic surfactant on its solubilising capacity is dependent on the location of the solubilisate within the micelle and is complicated by corresponding changes in the micellar size. The aggregation number decreases with increase in the hydrophilic chain length so there are more micelles in a given concentration of surfactant and, although the number of molecules solubilised per micelle decreases, the total amount solubilised per mole of surfactant may actually increase.

Nature of the solubilisate

- For a simple homologous series of solubilisates a decrease in solubilisation occurs when the alkyl chain length is increased.
- A relationship between the lipophilicity of the solubilisate, expressed by the partition coefficient between octanol and water, and its extent of solubilisation has been noted for several surfactant systems.

Temperature

- With most systems the amount solubilised increases as temperature increases.
- This increase is particularly pronounced with some non-ionic surfactants where it is a consequence of an increase in the micellar size with temperature increase.
- In some cases, although the amount of drug that can be taken up by a surfactant solution increases with temperature increase, this may simply reflect an increase in the amount of drug dissolved in the aqueous phase rather than an increased solubilisation by the micelles.

Pharmaceutical applications of solubilisation

- the solubilisation of phenolic compounds such as cresol, chlorocresol, chloroxylenol and thymol with soap to form clear solutions for use in disinfection

Tips

- Remember that the micelle core is like a tiny reservoir of hydrocarbon and it is therefore not surprising that there is a close relationship between the distribution of a compound between octanol and water phases in a test tube and its distribution between micelles and water in a micellar solution.
- A very lipophilic solubilisate will mainly reside in the micelles rather than in the aqueous phase surrounding them. This compound will therefore have a high micelle/water partition coefficient and also a high octanol/water partition coefficient.
- On the other hand a hydrophilic compound will be partitioned mainly in the aqueous phase rather than the micelles and will have a low micelle/water and octanol/water partition coefficient.

■   solubilised solutions of iodine in non-ionic surfactant micelles (iodophors) for use in instrument sterilisation

■   solubilisation of drugs (for example, steroids and water-insoluble vitamins), and essential oils by non-ionic surfactants (usually polysorbates or polyoxyethylene sorbitan esters of fatty acids).

Multiple choice questions

1.   Using Traube's rule, calculate the concentration of a surfactant with a hydrocarbon chain length of 16 carbon atoms that would be required to achieve the same lowering of the surface tension of water as a 8×10^{-4} mol dm^{-3} solution of a surfactant in the same homologous series with a hydrocarbon chain length of 18 carbon atoms:
   a. 0.89×10^{-4} mol dm^{-3}
   b. 7.2×10^{-3} mol dm^{-3}
   c. 2.4×10^{-3} mol dm^{-3}
   d. 4.8×10^{-3} mol dm^{-3}
   e. 1.33×10^{-4} mol dm^{-3}

2.   The slope of a plot of surface tension against logarithm of surfactant concentration for an aqueous surfactant solution measured at a concentration just below the CMC at a temperature of 30°C is –0.0115 N m^{-1}. Using the Gibbs equation, calculate the surface excess concentration in mol/m^2 given that $R =$ 8.314 J mol^{-1} K^{-1}:
   a. 1.98×10^{-6} mol m^{-2}
   b. 2.00×10^{-5} mol m^{-2}
   c. 5.04×10^{5} mol m^{-2}
   d. 1.64×10^{-5} mol m^{-2}
   e. 4.56×10^{-6} mol m^{-2}

3.   In relation to the surface tension at the air–water interface, indicate whether each of the following statements is *true* or *false*:
   a.   Surface tension is due to spontaneous expansion of the surface.
   b.   Surface tension arises because of the downward pull of molecules in the water.
   c.   Surface tension represents a state of maximum free energy.
   d.   Surface tension has units of N m^{-1}.
   e.   Surface tension is lowered by surface-active agents.

4.   A surfactant has a structure $CH_3(CH_2)_{11}(OCH_2CH_2)_8OH$. Indicate whether each of the following statements is *true* or *false*:
   a.   The surface activity of the surfactant will increase when the alkyl chain length is increased.
   b.   The CMC of the surfactant will increase when the alkyl chain length is increased.
   c.   The CMC of the surfactant will increase when the ethylene oxide chain length is decreased.

d. The hydrophobicity of the molecule will increase when the ethylene oxide chain length is increased.

e. Aqueous solutions of the surfactant will show a cloud point when heated.

5. **Indicate which of the following statements concerning the structure of micelles are correct:**

a. The core consists of the hydrophobic chains of the surfactant.

b. The Stern layer of ionic micelles contains the charged head groups.

c. In ionic micelles most of the counterions are contained in the Gouy–Chapman layer.

d. Micelles formed by non-ionic surfactants are generally much smaller than those formed by ionic surfactants with identical hydrophobic groups.

e. The Gouy–Chapman layer of an ionic micelle is compressed in the presence of electrolyte.

6. **In relation to the adsorption of an ionisable drug molecule onto an uncharged solid surface from an aqueous solution, indicate whether each of the following statements is *true* or *false*:**

a. The amount adsorbed usually increases as the ionisation of the drug decreases.

b. The amount adsorbed is not affected by change of pH.

c. The amount adsorbed usually decreases as the ionisation of the drug decreases.

d. The adsorptive capacity of the solid increases when its surface area is increased.

e. The adsorptive capacity of the solid is not affected by changes in its surface area.

7. **In relation to insoluble monolayers formed on the surface of water, indicate whether each of the following statements is *true* or *false*:**

a. Insoluble monolayers are formed by water-soluble surfactants.

b. The properties of insoluble monolayers are determined by an equilibrium between molecules in the monolayer and those in the bulk solution.

c. Polymers having several polar groups usually form gaseous films.

d. Molecules with bulky side chains usually form solid or condensed films.

e. The area occupied by a molecule in a gaseous film is greater than the cross-sectional area of the molecule.

8. **In relation to liquid crystals, indicate whether each of the following statements is *true* or *false*:**

a. Thermotropic liquid crystals are formed when concentrated surfactant solutions are heated.

b. In smectic liquid crystals the long axes of groups of molecules are parallel and organised in layers.

c. Variation of temperature and pressure of solutions of cholesteric liquid crystals can produce dramatic colour changes.

d. In the lamellar phase the surfactant molecules are arranged in bilayers.

e. The hexagonal phase is more fluid than the lamellar phase.

9. **Indicate which** *one* **of the following statements is correct.**
    The main reason why surfactants form micelles is because:
a. There is a decrease of entropy when surfactant molecules are transferred from water to a micelle.
b. There is an increase of entropy when surfactant molecules are transferred from water to a micelle.
c. There is a large decrease of enthalpy when micelles form.
d. There is a large increase of enthalpy when micelles form.
e. The free energy of the system increases when micelles form.

10. **Indicate which** *one* **of the following statements is correct.**
    In the solubilisation of poorly soluble drugs by aqueous surfactant solutions:
a. Non-polar drugs are usually solubilised in the palisade layer of a non-ionic micelle.
b. Polar drugs are usually solubilised in the micelle core.
c. Drugs with a high octanol/water partition coefficient will usually have a high micelle/water partition coefficient.
d. The solubilisation capacity of a non-ionic surfactant usually decreases with increase of temperature.
e. For a homologous series of solubilisates, an increase of solubilisation occurs when the alkyl chain length is increased.

chapter 5

Emulsions, suspensions and other dispersed systems

Overview

In this chapter we will:

- survey the variety of emulsions, suspensions and aerosols used in pharmacy
- discuss what contributes to their stability
- examine the elements of colloid stability theory and see how these assist the design of formulations.

Colloid stability

Water-insoluble drugs in fine dispersion form lyophobic dispersions. Because of their high surface energy they are thermodynamically unstable and have a tendency to aggregate.

Emulsions and aerosols are thermodynamically unstable two-phase systems which only reach equilibrium when the globules have *coalesced* to form one macro-phase, when the surface area is at a minimum.

Suspension particles achieve a lower surface area by *flocculating* or *aggregating*: they do not coalesce.

In dispersions of fine particles in a liquid (or of particles in a gas) frequent encounters between the particles occur due to:

- Brownian movement
- creaming or sedimentation
- convection.

According to Stokes' law the rate of sedimentation (or creaming), v, of a spherical particle in a fluid medium, viscosity η, is given by:

$$v = \frac{2ga^2(\rho_1 - \rho_2)}{9\eta}$$

KeyPoints

- Colloids can be broadly classified as:
- *lyophobic* (solvent-hating) (= *hydrophobic* in aqueous systems)
- *lyophilic* (solvent liking) (= *hydrophilic* in aqueous systems).
- Emulsions and suspensions are disperse systems – a liquid or solid phase dispersed in an external liquid phase.
- The *disperse phase* is the phase that is subdivided.
- The *continuous phase* is the phase in which the disperse phase is distributed.
- Emulsions and suspensions are intrinsically unstable systems that require stabilisers to ensure a useful lifetime.
- Emulsions exist in many forms:
- oil-in-water
- water-in-oil
- oil-in-oil (rare)
- a variety of multiple emulsions such as water-in-oil-in-water systems and oil-in-water-in-oil systems.
- Pharmaceutical emulsions and suspensions are in the colloidal state, i.e. the disperse phase sizes range from nanometres to the visible (several micrometres). Microemulsions are formulated so that the disperse phase is in the nanometre size range.
- Suspensions may have an aqueous or oily continuous phase.
- Aerosols are dispersions of a liquid or solid in air.

where the particle radius is a, ρ_1 is the density of the particles, ρ_2 is the density of the medium and g is the gravitational constant.

DLVO theory of colloid stability

Forces of interaction between colloidal particles

- van der Waals forces or electromagnetic forces (attraction)
- electrostatic forces (repulsion)
- Born forces – essentially short-range (repulsion)
- steric forces (repulsive) due to adsorbed molecules (particularly macromolecules) at the particle interface
- solvation forces (repulsive) due to reduction in the hydration of stabilising molecules on close approach.

Consideration of the electrostatic repulsion and van der Waals forces of attraction by Deryagin, Landau, Verwey and Overbeek (DLVO) led to a theory of stability of hydrophobic suspensions.

DLVO theory considers two spherical particles of radius a at a distance apart H (Figure 5.1).

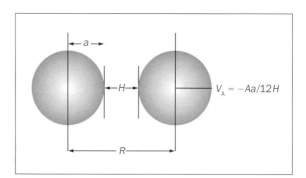

Figure 5.1 Diagram of the interaction between two spheres of radius a at a distance of separation H with a centre-to-centre distance of $R = H + 2a$.

In this theory:
- The combination of the electrostatic repulsive energy (V_R) with the attractive potential energy (V_A) gives the total potential energy of interaction:

$$V_{total} = V_A + V_R$$

- *Attractive forces* arise from van der Waals forces between particles of the same kind.
- When the particles are large relative to the distance of separation, the attractive force (V_A) is written as:

$$V_A = -\frac{Aa}{12H}$$

where A is the Hamaker constant.

- *Repulsive forces* arise from the electrical charge on particles, which is due either to ionisation of surface groups or to adsorption of ions:
- A particle surface with a negative charge has a layer of positive ions attracted to its surface in the Stern layer, and a *diffuse* or *electrical double layer* which accumulates and contains both positive and negative ions (Figure 5.2).

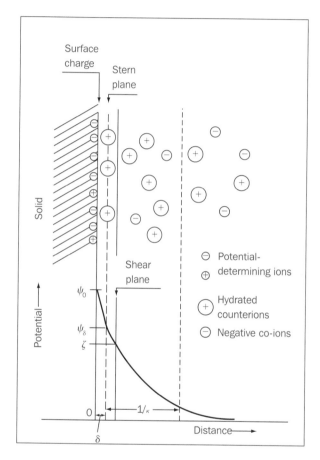

Figure 5.2 Distribution of charges at the surface of a negatively charged solid.

- Electrostatic forces arise from the interaction of these electrical double layers surrounding particles in suspension, leading to repulsion if the particles have both the same positive or negative surface charges.
- The electrostatic repulsive force decays as an exponential function of the distance. It has a range of the order of the thickness of the electrical double layer, equal to the Debye–Hückel length, $1/\kappa$. An approximate equation for the repulsive

interactions for small surface potentials and low values of κ is:

$$V_R = 2\pi\varepsilon\varepsilon_0 a\psi_\delta^2 \exp(-\kappa H)$$

where ε_0 is the permittivity of a vacuum, ε is the dielectric constant (or relative permittivity) of the dispersion medium and ψ_δ is the Stern potential, which can be approximated to the zeta potential (ζ) measured by microelectrophoresis.

■ V_{total} plotted against the distance of separation H gives a potential energy curve (Figure 5.3) showing maximum and minimum energy states.

Figure 5.3 Typical DLVO plot.

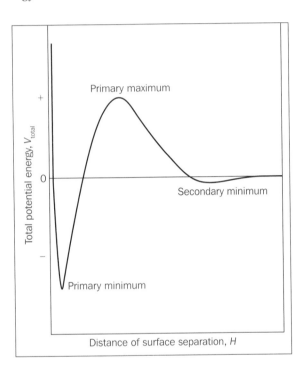

KeyPoints

- If the primary maximum is too small, two interacting particles may reach the primary minimum and the depth of this energy minimum means escape is improbable.
- When the primary maximum is sufficiently high, the two particles do not reach the stage of being in close contact.
- The depth of the secondary minimum is important in determining the stability of the system.
- If the secondary minimum is less than the thermal energy, kT (where k is the Boltzmann constant), the particles will always repel each other.

Effect of electrolytes on stability

Figure 5.4 shows the effect of electrolyte on a typical DLVO plot. Changes in the plot arise because of compression of the double layer as the electrolyte concentration is increased, which increases κ, so decreasing $1/\kappa$.

■ At low electrolyte concentrations the range of the double layer is high and V_R extends to large distances around the particles. Summation of V_R and V_A gives a total energy curve having a high primary maximum but no secondary minimum.

- The decrease of the double layer when more electrolyte is added produces a more rapid decay in V_R and the result is a small primary maximum but, more importantly, a secondary minimum. This concentration of electrolyte would produce a stable suspension, since flocculation could occur in the secondary minimum. The small primary maximum would be sufficient to prevent coagulation in the primary minimum.
- At high concentrations of added electrolyte, the range of V_R would be so small that the van der Waals attractive forces alone dictate the shape of the energy curve. The curve has no primary maximum or secondary minimum.

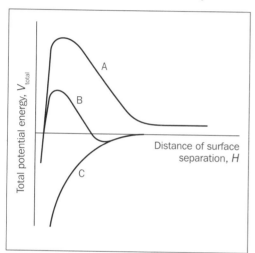

Figure 5.4 The effect on the DLVO plot of
A: low
B: medium and
C: high concentrations of added electrolyte.

The magnitude of the effect of an electrolyte of a given concentration on V_R also depends on the valence of the ion of opposite charge to that of the particles (the counterion): the greater the valence of the added counterion, the greater its effect on V_R. These generalisations are known as the *Schulze–Hardy rule*. Notice that it does not matter which particular counterion of a given valence is added.

Repulsion between hydrated surfaces – steric stabilisation

The DLVO theory deals only with electrostatic repulsion whereas colloids can also be stabilised by the repulsive forces that arise from adsorption of macromolecules and surfactants to their surfaces. In aqueous media these adsorbed molecules will be hydrated.

Stabilisation arises because of three effects:
1. Entropic effect:
- Loss of freedom of movement of the chains of the adsorbed molecules

- the approach of two particles with adsorbed stabilising chains leads to a steric interaction when the chains interact.
- The steric effect does not come into play until $H = 2\delta$, so the interaction increases suddenly with decreasing distance.
- The loss of conformational freedom leads to a negative entropy change ($-\Delta S$). Each chain loses some of its conformational freedom and its contribution to the free energy of the system is increased, leading to the repulsion.
- The steric effect depends on: (1) the chain length of the adsorbed molecule; (2) the interactions of the solvent with the chains; and (3) the number of chains per unit area of interacting surface.

2. Osmotic effect:

- The 'osmotic effect' arises as the macromolecular chains on neighbouring particles crowd into each other's space, increasing the concentration of chains in the overlap region.
- The repulsion which arises is due to the osmotic pressure of the solvent attempting to dilute out the concentrated region: this can only be achieved by the particles moving apart.

3. Enthalpic stabilisation:

- On close approach of the particles the hydrating water on the adsorbed molecules is released, which causes an increase in enthalpy leading to repulsion (Figure 5.5).

Figure 5.5 Representation of enthalpic stabilisation of particles with adsorbed hydrophilic chains.

Tips

In considering the reasons for the stabilising action of the adsorbed molecules the basic thermodynamic equation must be remembered:

$$\Delta G = \Delta H - T\Delta S$$

ΔG is the change in free energy of a process such as repulsion: effective repulsion of the particles is characterised by an increase of ΔG.

ΔH is the change in enthalpy: loss of water molecules leads to a positive change in enthalpy, hence an increase of free energy.

ΔS is the change in entropy: loss of freedom of movement of stabilising macromolecules leads to a negative entropy change, hence positive ΔG.

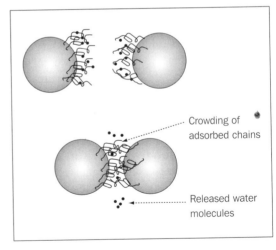

Crowding of adsorbed chains

Released water molecules

Emulsions

Stability of oil-in-water and water-in-oil emulsions

Adsorption of a surfactant at the oil–water interface lowers interfacial tension, hence aids the dispersal of the oil into droplets of a small size and helps to maintain the particles in a dispersed state (Figure 5.6).

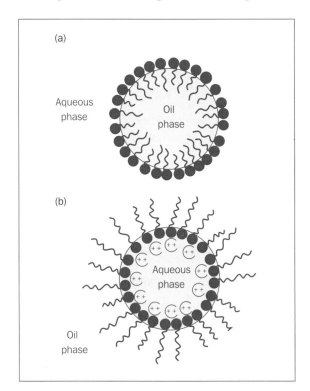

Figure 5.6 Surfactant films at the oil–water interface for the stabilisation of (a) oil-in-water and (b) water-in oil emulsions.

Unless the interfacial tension is zero, there is a natural tendency for the oil droplets to coalesce to reduce the area of oil–water contact, but the presence of the surfactant monolayer at the surface of the droplet reduces the possibility of collisions leading to coalescence.

Adsorption of charged surfactants will lead to an increase in zeta potential and will thus help to maintain stability by increasing V_R.

Non-ionic surfactants such as the alkyl or aryl polyoxyethylene ethers or polyoxyethylene-polyoxypropylene-polyoxyethylene block copolymers are widely used in pharmaceutical emulsions. These adsorb onto the emulsion droplets and maintain stability by creating a hydrated layer on the hydrophobic particle in oil-in-water emulsions.

In water-in-oil emulsions the hydrocarbon chains of the adsorbed molecules protrude into the oily continuous phase. Stabilisation arises from steric repulsive forces.

It is usually observed that mixtures of surfactants form more stable emulsions than do single surfactants, perhaps due to complex formation at the interface providing a more 'rigid' stabilising film.

HLB system

The hydrophile–lipophile balance (HLB) number is a measure of the balance between hydrophobic and hydrophilic portions of a surfactant.

In selecting a surfactant for emulsion stabilisation it is essential that there is a degree of surfactant hydrophilicity to confer an enthalpic stabilising force and a degree of hydrophobicity to secure adsorption at the oil-in-water interface.

The HLB of a surfactant is expressed using an arbitrary scale which for non-ionic surfactants ranges from 0 to 20.

- At the higher end of the scale, the surfactants are hydrophilic and act as solubilising agents, detergents and oil-in-water emulsifiers.
- Oil-soluble surfactants with a low HLB act as water-in-oil emulsifiers.

HLB values can be calculated according to empirical but useful formulae:

- For simple alkyl ethers in which the hydrophile consists only of ethylene oxide,

$$HLB = E/5$$

where E is the weight percentage of ethylene oxide groups.
- The HLB of polyhydric alcohol fatty acid esters such as glyceryl monostearate may be obtained from the equation:

$$HLB = 20\left(1 - \frac{S}{A}\right)$$

where S is the saponification number of the ester and A is the acid number of the fatty acid. The HLB of polysorbate 20 (Tween 20) calculated using this formula is 16.7, S being 45.5 and $A = 276$. The polysorbate (Tween) surfactants have HLB values in the range 9.6–16.7; the sorbitan ester (Span) surfactants have HLBs in the lower range of 1.8–8.6.
- For those materials for which it is not possible to obtain saponification numbers, for example beeswax and lanolin derivatives, the HLB is calculated from:

$$HLB = (E + P)/5$$

where P is the weight percentage of polyhydric alcohol groups (glycerol or sorbitol) in the molecule.
- HLB values can also be calculated from group contributions using:

$$HLB = \Sigma(\text{hydrophilic group numbers}) - \Sigma(\text{lipophilic group numbers}) + 7$$

- For a mixture of two surfactants containing fraction f of A and $(1 - f)$ of B it is assumed that the HLB is an algebraic mean of the two HLB numbers:

$$HLB_{mixture} = fHLB_A + (1 - f)HLB_B$$

The HLB system has several drawbacks:
- The calculated HLB cannot take account of the effect of temperature or that of additives.
- The presence of agents which salt-in or salt-out surfactants will, respectively, increase and decrease the effective (as opposed to the calculated) HLB values. Salting-out the surfactant (e.g. with NaCl) will make the molecules more hydrophobic (less hydrophilic).

Choice of emulsifier or emulsifier mixture
- The appropriate choice of emulsifier or emulsifier mixture can be made by preparing a series of emulsions with a range of surfactants of varying HLB.
- Mixtures of surfactants with high HLB and low HLB give more stable emulsions than do single surfactants.
- The solubility of surfactant components in both the disperse and the continuous phase maintains the stability of the surfactant film at the interface from the reservoir created in each phase.
- In the experimental determination of optimum HLB the system with the minimum creaming or separation of phases is deemed to have an optimal HLB. It is therefore possible to determine optimum HLB numbers required to produce stable emulsions of a variety of oils.
- At the optimum HLB the mean particle size of the emulsion is at a minimum, which explains the increased stability.
- The formation of a viscous network of surfactants in the continuous phase prevents their collision and this effect overrides the influence of the interfacial layer and barrier forces due to the presence of adsorbed layers.

Multiple emulsions
- Multiple emulsions are emulsions whose disperse phase contains droplets of another phase.
- They are made by emulsifying a water-in-oil emulsion with

Tips

- The *saponification number* of the ester represents the number of milligrams of alkali (NaOH or KOH) required to hydrolyse 1 g of the ester. When the ester is an oil or fat the esterification leads to the formation of a soap, which is why the esterification reaction is called 'saponification'. As most of the mass of a oil or fat is in the fatty acids, it is a measure of the average molecular weight (or chain length) of all the fatty acids present.
- The *acid number* is the number of milligrams of KOH required to neutralize 1 g of the fatty acid. The acid number is a measure of the number of carboxylic acid groups in the fatty acid.

a hydrophilic surfactant to produce a water-in-oil-in-water system, or an oil-in-water system with a low HLB surfactant to produce an oil-in-water-in-oil system. Other forms can be made.

- Water-in-oil emulsions in which a water-soluble drug is dissolved in the aqueous phase may be injected by the subcutaneous or intramuscular routes to produce a delayed-action preparation. To escape, the drug has to diffuse through the oil to reach the tissue fluids, hence the delayed-release action.

- The main disadvantage of a water-in-oil emulsion is its high viscosity because of the oil continuous phase. Emulsifying a water-in-oil emulsion using surfactants which stabilise an oily disperse phase can produce multiple water-in-oil-in-water emulsions with an external aqueous phase and lower viscosity than the primary emulsion.

- Physical degradation of water-in-oil-in-water emulsions can arise by several routes:
  - coalescence of the internal water droplets
  - coalescence of the oil droplets surrounding them
  - rupture of the oil film separating the internal and external aqueous phases
  - osmotic flux of water to and from the internal droplets, possibly associated with inverse micellar species in the oil phase.

Microemulsions

- Microemulsions are homogeneous transparent systems of low viscosity which contain a high percentage of both oil and water and high concentrations (15–25%) of emulsifier mixture.

- Microemulsions form spontaneously when the components are mixed in the appropriate ratios and are thermodynamically stable.

- In their simplest form, microemulsions are small droplets (diameter 5–140 nm) of one liquid dispersed throughout another. The droplet size is therefore very much smaller than that of normal emulsions (which is why microemulsions are transparent) and the droplets are very much more uniform in size.

- They can be dispersions of oil droplets in water or water droplets in oil but more complex structures (bicontinuous structures) may exist when there are almost equal amounts of oil and water.

- An essential requirement for their formation and stability is the attainment of a very low interfacial tension γ. Since microemulsions have a very large interface between oil and water (because of the small droplet size), they can only be thermodynamically stable if the interfacial tension is so low that the positive interfacial energy (given by γA, where A is the interfacial area) can be compensated by the negative free energy of mixing (ΔG_m).

■ To achieve the very low interfacial tension required for their formation it is usually necessary to include a second amphiphile (the *cosurfactant*) such as a short-chain alcohol in the formulation. This cosurfactant is incorporated into the interfacial film around the droplets (Figure 5.7).

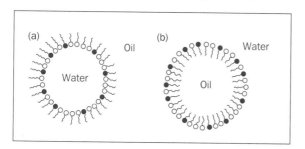

Figure 5.7 Diagrammatic representation of the interfacial layers around microemulsion droplets in (a) water-in-oil and (b) oil-in-water microemulsions.

Semi-solid emulsions (creams, ointments)

Stable oil-in-water creams prepared with ionic or non-ionic emulsifying waxes are composed of (at least) four phases:

1. a dispersed oil phase
2. a crystalline gel phase
3. a crystalline hydrate phase
4. a bulk aqueous phase containing a dilute solution of surfactant.

The interaction of the surfactant and fatty alcohol components of emulsifying mixtures which leads to high viscosity (body) is time-dependent, giving the name 'self-bodying' to these emulsions.

The overall stability of a cream is dependent on the stability of the crystalline gel phase.

The liquid crystalline phases form multilayers at the oil–water interface. These protect against coalescence by reducing the van der Waals forces of attraction and by retarding film thinning between approaching droplets; the viscosity of the liquid crystalline phases is often 100 times greater than that of phases without these structures.

Biopharmaceutical aspects of emulsions

■ Traditionally emulsions were used to administer oils such as castor oil and liquid paraffin in a palatable form. This is now a minor use.

■ Emulsions are of interest as vehicles for drug delivery in which the drug is dissolved in the disperse phase. For example, lipid oil-in-water emulsions are used as vehicles for lipophilic drugs (diazepam, propofol) for intravenous use.

■ Griseofulvin and indoxole in emulsion formulations exhibit enhanced oral absorption.

Preservative availability in emulsified systems

■ Microbial spoilage of emulsified products is avoided by the inclusion of appropriate amounts of a preservative in the formulation.

■ Preservatives in emulsions may partition to the oily or micellar phases of complex systems and some are inactivated by surfactants, hence calculations must be made of the appropriate amounts.

■ The presence of surfactant micelles alters the native partition coefficient of the preservative molecule because the micellar phase offers an alternative site for preservative molecules. Partitioning then occurs between the oil globule and the aqueous micellar phases, decreasing the amounts in the aqueous phase, where they are active.

Intravenous fat emulsions

■ Cottonseed oil or soybean oil emulsions are used to supply a large amount of energy in a small volume of isotonic liquid; they supply the body with essential fatty acids and triglycerides.

■ Fat emulsions for intravenous nutrition contain vegetable oil and phospholipid emulsifier.

■ Several commercial fat emulsions are available, such as Intralipid, Lipiphysan, Lipofundin and Lipofundin S. They contain either cottonseed oil or soybean oil. Purified egg-yolk phospholipids are used as the emulsifiers in Intralipid.

■ Isotonicity is obtained by the addition of sorbitol, xylitol or glycerol.

■ Intralipid has also been used as the basis of an intravenous drug carrier, for example for diazepam (Diazemuls) and propofol (Diprivan).

■ The addition of electrolyte or drugs to intravenous fat emulsions is generally contraindicated because of the risk of destabilising the emulsion.

The rheology of emulsions

■ Most emulsions display both plastic and pseudoplastic flow behaviour rather than simple Newtonian flow.

■ The pourability, spreadability and 'syringeability' of an emulsion will, however, be directly determined by its rheological properties.

■ The high viscosity of water-in-oil emulsions leads to problems with intramuscular administration of injectable formulations. Conversion to a multiple emulsion (water-in-oil-in-water) leads to a dramatic decrease in viscosity and consequent improved ease of injection.

Suspensions

Stability of suspensions

Flocculation of suspensions

■ In *deflocculated systems* the particles are not associated; pressure on the individual particles can lead to close packing of the particles at the bottom of the container to such an extent that the secondary energy barriers are overcome and the particles are forced together in the primary minimum of the DLVO plot and become irreversibly bound together to form a cake.

■ Caking of the suspension is usually prevented by including a flocculating agent in the formulation: it cannot be eliminated by reduction of particle size or by increasing the viscosity of the continuous phase.

■ In *flocculated systems* (where the repulsive barriers have been reduced) particles form loosely bonded structures (flocs or flocculates) in the secondary minimum of the DLVO plot. The particles therefore settle as flocs and not as individual particles (Figure 5.8). Because of the random arrangement of the particles in the flocs, the sediment is not closely packed and caking does not readily occur. Clearance of the supernatant is, however, too rapid for an acceptable pharmaceutical formulation.

■ The aim in the formulation of suspensions is, therefore, to achieve *partial or controlled flocculation*.

Suspension stability may be assessed by measurement of:

1. The ratio R of sedimentation layer volume (V_s) to total suspension volume (V_t). The height of the sedimented layer

> **KeyPoints**
>
> ■ An acceptable suspension has the following characteristics:
> – Suspended material should not settle too rapidly.
> – Particles that settle to the bottom of the container should not form a hard mass (cake) but should be easily redispersed on shaking.
> ■ The suspension should not be too viscous to pour freely from a bottle or to flow through a needle.
> ■ When a drug is suspended in a liquid, however, sedimentation, caking (leading to difficulty in resuspension), flocculation and particle growth (through dissolution and recrystallisation) can all occur.
> ■ Formulation of pharmaceutical suspensions to minimise caking can be achieved by the production of flocculated systems.
> ■ A flocculate or floc is a loose open structure or cluster of particles. A suspension consisting of particles in this state is termed flocculated; there are various states of flocculation and deflocculation.
> ■ Flocculated systems clear rapidly and the preparation often appears unsightly, so a partially deflocculated formulation is more acceptable.

(h_∞) is usually measured against the height of the suspension (h_0) so that:

$$R = \frac{V_s}{V_t} \approx \frac{h_\infty}{h_0}$$

Figure 5.8 Sedimentation of (a) deflocculated and (b) flocculated suspensions.

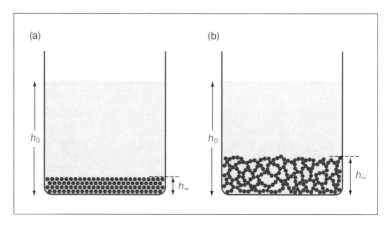

2. The zeta potential of the suspension particles:
   – Most suspension particles dispersed in water have a charge acquired by specific adsorption of ions or by ionisation of ionisable surface groups. If the charge arises from ionisation, the charge on the particle will depend on the pH of the environment.
   – Repulsive forces arise because of the interaction of the electrical double layers on adjacent particles.
   – The magnitude of the charge can be determined by measurement of the electrophoretic mobility of the particles in an electrical field. The velocity of migration of the particles (μ_E) under unit-applied potential can be determined microscopically with a timing device and eye-piece graticule.
   – For non-conducting particles, the Henry equation is used to obtain ζ from μ_E. This equation can be written in the form:

$$\mu_E = \frac{\zeta \varepsilon}{4\pi\eta} f(\kappa a)$$

where $f(\kappa a)$ varies between 1, for small κa, and 1.5, for large κa; ε is the dielectric constant of the continuous phase and η is its viscosity. In systems with low values of κa the equation can be written in the form:

$$\mu_E = \frac{\zeta \varepsilon}{4\pi\eta}$$

- The zeta potential (ζ) is not the surface potential (ψ_o) as discussed earlier but is related to it. ζ can be used as a reliable guide to the magnitude of electric repulsive forces between particles. Changes in ζ on the addition of flocculating agents, surfactants and other additives can then be used to predict the stability of the system.

Controlled flocculation

- In suspensions of charged particles the flocculation may be controlled by the addition of electrolyte or ionic surfactants that reduce the zeta potential, and hence V_R, to give a satisfactory secondary minimum in which flocs may be formed. Figure 5.9 shows the changes in a bismuth subnitrate suspension on addition of dibasic potassium phosphate as flocculating agent.
- In the absence of charge on the particles flocculation may be controlled using non-ionic polymeric material including naturally occurring gums (e.g. tragacanth) and cellulose polymers (e.g. sodium carboxymethylcellulose). These polymers increase the viscosity of the aqueous vehicle, so hindering the movement of the particles, and also may form adsorbed layers on the particles which influence stability through steric stabilisation and, in some cases, bridging between particles.
- The ideal suspending agent for controlling flocculation should:
- be readily and uniformly incorporated in the formulation
- be readily dissolved or dispersed in water without resort to special techniques
- ensure the formation of a loosely packed system which does not cake
- not influence the dissolution rate or absorption rate of the drug
- be inert, non-toxic and free from incompatibilities.

Non-aqueous suspensions

- Many pharmaceutical aerosols consist of solids dispersed in a non-aqueous propellant or propellant mixture.
- Low amounts of water adsorb at the particle surface and can lead to aggregation of the particles or to deposition on the walls of the container, which adversely affects the product.

Adhesion of suspension particles to containers

- When the walls of a container are wetted repeatedly an adhering layer of suspension particles may build up, and this subsequently dries to a hard and thick layer.
- Where the suspension is in constant contact with the container wall, immersional wetting occurs, in which particles are pressed up to the wall and may or may not adhere.

Figure 5.9 Controlled flocculation of a bismuth subnitrate suspension using dibasic potassium phosphate (KH_2PO_4) as the flocculating agent.

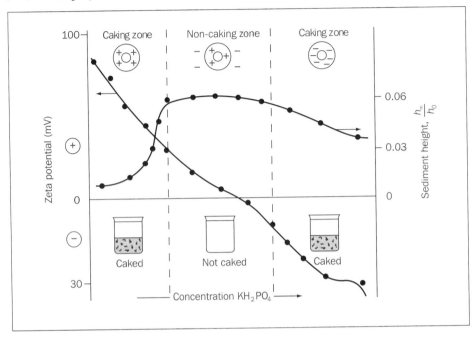

- Above the liquid line, spreading of the suspension during shaking or pouring may also lead to adhesion of the particles contained in the spreading liquid.
- Adhesion increases with increase in suspension concentration, and with the number of contacts the suspension makes with the surfaces.

Foams and defoamers

- Aqueous foams are formed from a three-dimensional network of surfactant films in air.
- Foams can be used as formulations for the delivery of enemas and topical drugs.
- Foams which develop in the production of liquids are troublesome, hence there is an interest in breaking foams and preventing foam formation. Small quantities of specific agents can reduce foam stability markedly. There are two types of such agent:
  1. *Foam breakers*, which are thought to act as small droplets forming in the foam lamellae.
  2. *Foam preventives*, which are thought to adsorb at the air–water interface in preference to the surfactants which stabilise the thin films.

■ The most important action of an antifoam agent is to eliminate surface elasticity, the property that is responsible for the durability of foams. To do this the antifoam agent:
- must displace any foam stabiliser
- must therefore have a low interfacial tension in the pure state to allow it to spread when applied to the foam.

Many foams can be made to collapse by applying drops of liquids such as ether, or long-chain alcohols such as octanol. Silcone fluids which have surface tensions as low as 20 mN m^{-1} are effective and more versatile than soluble antifoams.

Multiple choice questions

1. According to Stokes' law, which of the following changes to a formulation of an oil-in-water emulsion would be expected to decrease the rate of creaming of the emulsion?
   a. decrease in the size of the oil droplets
   b. increase in the viscosity of the continuous phase
   c. increase in the difference in density between the oil and water phases

2. Which of the following lead to attractive interaction between two particles?
   a. Born forces
   b. electrostatic forces
   c. van der Waals forces
   d. steric forces
   e. solvation forces

3. Indicate which of the following statements is true. Two particles will repel each other when:
   a. The primary maximum is very small.
   b. The secondary minimum is less than the thermal energy.
   c. The primary minimum is very deep.

4. Indicate which of the following statements are true. When electrolyte is added to a colloidal dispersion:
   a. The double-layer thickness around the particles is increased.
   b. Repulsion between particles is decreased.
   c. van der Waals forces between particles are decreased.
   d. The height of the primary maximum is decreased.

5. **Indicate which of the following statements are true. Repulsion between hydrated surfaces:**
a. results from an increase in the freedom of movement of the chains of the adsorbed molecules
b. increases with increase of chain length of the adsorbed molecules
c. results from release of hydrated water molecules from the chains of the adsorbed molecules
d. increases with an increase of the number of chains per unit area of interacting surface

6. **Indicate which of the following statements are true. Stabilisation of oil-in-water emulsions by surfactants:**
a. arises because of a reduction of the oil–water interfacial tension
b. is a consequence of a decrease of the zeta potential of the oil droplets
c. is usually more effective when more than one surfactant is used
d. can only be achieved with ionic surfactants

7. **Indicate which of the following statements is true. Oil-soluble surfactants:**
a. have high HLB values
b. are hydrophilic
c. can be used as emulsifiers to produce water-in-oil emulsions
d. are efficient solubilising agents

8. **Which of the following properties are characteristic of microemulsions?**
a. high surfactant content
b. droplet size greater than 1 μm
c. transparent systems
d. thermodynamically stable

9. **Which of the following properties are characteristic of deflocculated suspensions?**
a. close packing of the sediment to form a cake
b. slow sedimentation rate
c. formation of flocs
d. rapid clearance of supernatant

10. **A polysorbate has a molecular weight of 1300, an ethylene oxide weight percentage of 68 and a sorbitol weight percentage of 14. The HLB of this surfactant is:**
a. 15.0
b. 13.6
c. 16.4
d. 2.8

chapter 6
Polymers

Overview

In this chapter we will:

- look at the variety of structures formed by polymers and the properties of polymers in solution
- examine the solution properties and gelation of polymers and the characteristic properties of polymer gels
- look at the structure and properties of some typical polymers used in pharmacy and medicine
- discuss some of the many applications of polymers in the fabrication of drug delivery devices.

Polymer structure

Polymers consist of a large number of monomer units linked together in a long chain:

- For example, polyethylene is composed of repeating ethylene monomers:
  $$CH_2=CH_2 \rightarrow -CH_2-CH_2-CH_2-CH_2$$

Polymers in which all the monomer units are identical are referred to as homopolymers:

- Examples include polystyrene, polyethylene, poly(vinyl alcohol), polyacrylamide and polyvinylpyrrolidone.
- There are usually between about 100 and 10 000 monomer units in a chain.
- There may also be homopolymers with much smaller chains including *dimers* (2 monomer units), *trimers* (3 units) and *tetramers* (4 units). These small chains are called *oligomers*.

Side chains or substituents (R) may be attached to the repeating monomer units, as for example in vinyl polymers of the type $-H_2C-CH(R)-$. These may have:

- all the R groups on the same side of the polymer backbone (*isotactic*)
- a regular alternation of the R groups above and below the backbone (*syndiotactic*)
- a random arrangement of R groups above and below the backbone (*atactic*) (Figure 6.1).

KeyPoints

- Polymers are substances of high molecular weight made up of repeating *monomer* units.
- Polymer molecules may be linear or branched, and separate linear or branched chains may be joined by cross-links. Extensive cross-linking leads to a three-dimensional and often insoluble polymer network.
- Polymers in which all the monomeric units are identical are referred to as *homopolymers*; those formed from more than one monomer type are called *copolymers*.

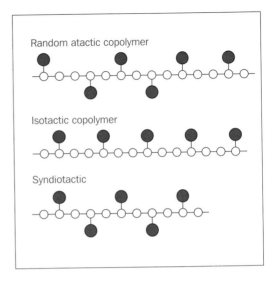

Figure 6.1 Representation of isotactic, syndiotactic and atactic polymers showing the position of the substituent group ● in relation to the polymer backbone —○–○–○–○–.

Polymers formed from more than one type of monomer are referred to as copolymers. There are several different types of copolymer (Figure 6.2):

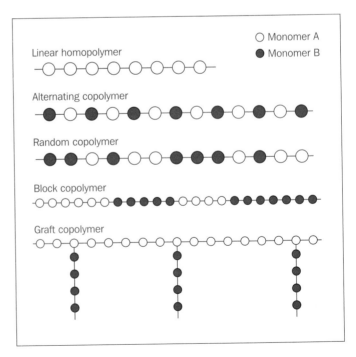

Figure 6.2 Varieties of copolymer structure attainable through the polymerisation of two different monomers represented by ● and ○.

■ The different monomers can be arranged in a linear chain in either a *random* manner or in an *alternating* pattern along the chain.

■ The linear polymer chains may be constructed from blocks of each monomer and are then referred to as *block* copolymers. These may be:

– *Diblock* copolymers in which there is a block composed of monomer A chains attached to a block of monomer B chains (AB diblocks).

– *Triblock* copolymers which are composed of either a block of A chains attached to a block of B chains attached to a block of A chains (ABA triblocks) or alternatively arranged as BAB triblocks. Well-known examples of ABA triblock copolymers are *poloxamers* in which the A chains are polyoxyethylene and the B chains are polyoxypropylene:

i.e. $HO(CH_2CH_2O)_x (CH(CH_3)CH_2O)_y (CH_2CH_2O)_x H$

x and y denote the numbers of monomer units in each of the blocks.

■ The chains may be composed of a backbone of repeat units of one monomer on to which is grafted chains of the second monomer in a comb-like manner. These copolymers are called *graft* copolymers.

Polymer chains can be linear (forming random coils in solution) or branched. There may be cross-linking between chains to form three-dimensional networks. Highly branched polymers (*dendrimers*) built around a central core can be synthesised with a range of sizes depending on the generation of the dendrimer.

Polymers do not form perfect crystals but have crystalline regions surrounded by amorphous regions (Figure 6.3).

■ The melting point of polymers is not as well defined as in low-molecular-weight crystalline solids because of the presence of the poorly structured regions which melt over a range of temperatures.

■ As well as the melting point the polymer may also exhibit a *glass transition temperature*, T_g. Below T_g the chains are 'frozen' in position and the polymer is glassy and brittle; above T_g the chains are mobile and the polymer is tougher and more flexible.

Figure 6.3 Diagrammatic representation of a solid polymer showing regions of crystallinity and regions which are amorphous.

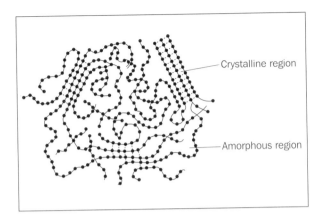

Crystalline region

Amorphous region

Solution properties of polymers

Polydispersity

Nearly all synthetic polymers and naturally occurring macromolecules possess a range of molecular weights. The exceptions to this are proteins and natural polypeptides. The molecular weight is thus an average molecular weight and depending on the experimental method used to measure it may be:

- a *number average* molecular weight, M_n, (determined by chemical analysis or osmotic pressure measurement) which, in a mixture containing n_1, n_2, n_3... moles of polymer with molecular weights M_1, M_2, M_3..., respectively, is defined by:

$$M_n = \frac{n_1 M_1 + n_2 M_2 + n_3 M_3 + \ldots\ldots}{n_1 + n_2 + n_3 + \ldots\ldots} = \frac{\Sigma n_i M_i}{\Sigma n_i}$$

- a *weight average* molecular weight, M_w (determined by light scattering methods):

$$M_w = \frac{m_1 M_1 + m_2 M_2 + m_3 M_3 + \ldots\ldots}{m_1 + m_2 + m_3 + \ldots\ldots} = \frac{\Sigma n_i M_i^2}{\Sigma n_i M_i}$$

where m_2, m_3...... are the masses of each species.

The *weight average* molecular weight, M_w, is biased towards large molecules and in a polydisperse polymer is always greater than M_n. The ratio M_w/M_n expresses the degree of polydispersity.

KeyPoints

- There is usually a range of sizes of the polymer chains in solution, i.e. the polymer solutions are polydisperse.
- As a consequence of polydispersity the measured molecular weight varies depending on the experimental method used – some techniques such as light scattering are more influenced by the larger molecules than others such as osmometry.
- The viscosity of a polymer solution not only depends on its concentration but also on polymer–solvent interactions, charge interactions and the binding of small molecules.

Tip

Remember that the mass, m_i, of a particular species is obtained by multiplying the molecular weight of each species by the number of molecules of that weight; that is, $m_i = n_i M_i$. Thus the molecular weight appears as the square in the numerator of the equation for the weight average molecular weight.

Viscosity

Assuming that the polymer solution exhibits Newtonian flow properties, the viscosity can be expressed by:

- The *relative viscosity*, η_{rel}, defined as the ratio of the viscosity of the solution, η, to the viscosity of the pure solvent η_0, i.e, $\eta_{rel} = \eta/\eta_0$.
- The *specific viscosity*, η_{sp}, of the solution defined by $\eta_{sp} = \eta_{rel} - 1$.
- The *intrinsic viscosity* $[\eta]$ obtained by extrapolation of plots of the ratio η_{sp}/c (called the *reduced viscosity*) against concentration c to zero concentration.

If the polymer forms spherical particles in dilute solution then $\eta_{rel} = 1 + 2.5\phi$ where ϕ is the volume fraction (volume of the particles divided by the total volume of the solution).

- Therefore, a plot of η_{sp}/ϕ against ϕ should have an intercept of 2.5.
- Departure of the limiting value from this theoretical value may result from either hydration of the particles, or particle asymmetry, or both.

A change in shape due to changes in polymer–solvent interactions (Figure 6.4) and the binding of small molecules with the polymer may lead to significant changes in solution viscosity:

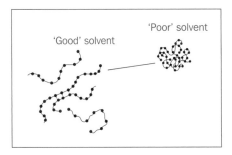

Figure 6.4 Representation of the conformation of a polymer in 'good' and 'poor' solvents.

- In 'good' solvents linear macromolecules will be expanded as the polar groups will be solvated.
- In 'poor' solvents the intramolecular attraction between the segments is greater than the segment–solvent affinity and the molecule will tend to coil up.

The viscosity of ionised polymers is complicated by charge interactions which vary with polymer concentration and additive concentration.

KeyPoints

- A gel is a polymer–solvent system containing a three-dimensional network which can be formed by swelling of solid polymer or by reduction in the solubility of the polymer in the solution.
- When gels are formed from solutions, each system is characterised by a critical concentration of gelation below which a gel is not formed.
- Gelation is characterised by a large increase in viscosity above the gel point, the appearance of a rubber-like elasticity and a yield point stress at higher polymer concentrations.
- Gels can be irreversible or reversible systems depending on the nature of the bonds between the chains of the network.

- Flexible, charged macromolecules will vary in shape with the degree of ionisation. At maximum ionisation they are stretched out due to mutual charge repulsion and the viscosity increases. On addition of small counterions the effective charge is reduced and the molecules contract; the viscosity falls as a result.

The intrinsic viscosity of solutions of linear high-molecular-weight polymers is proportional to the molecular weight M of the polymer as given by the *Staudinger equation*:

$$[\eta] = KM^a$$

where a is a constant in the range 0–2 (for most high polymers a has a value between 0.6 and 0.8) and K is a constant for a given polymer–solvent system.

Properties of polymer gels

Gels can be divided into two groups, depending on the nature of the bonds between the chains of the network:

- Gels of *type I* are irreversible systems with a three-dimensional network formed by covalent bonds between the macromolecules. They include swollen networks that have been formed by polymerisation of a monomer in the presence of a cross-linking agent:
  - Examples include poly(hydroxyethyl methacrylate) (poly[HEMA]) which may be cross-linked with, for example, ethylene glycol dimethacrylate (EGDMA).
  - These polymers swell in water but cannot dissolve as the cross-links are stable.
  - This expansion on contact with water has been put to many uses, such as in the fabrication of expanding implants from cross-linked hydrophilic polymers which imbibe body fluids and swell to a predetermined volume.
  - Hydrophilic contact lenses (such as Soflens) are made from cross-linked poly[HEMA]s.
- *Type II* gels are heat-reversible, being held together by intermolecular bonds such as hydrogen bonds:
  - The temperature at which gelation occurs is called the *gel point* and gelation can be induced either by cooling (e.g. poly(vinyl alcohol)) or heating (e.g. water-soluble methylcelluloses) to this temperature depending on the type of temperature variation of solubility.

- The gel point can be influenced by the presence of additives which can induce gel formation by acting as bridge molecules (for example, with borax and poly(vinyl alcohol)) or by the addition of solvents such as glycerol.
- Because of their gelling properties poly(vinyl alcohol)s are used for application of drugs to the skin; on application the gel dries rapidly leaving a plastic film with the drug in intimate contact with the skin.
- Solutions of some poly(oxyethylene)-poly(oxypropylene)-poly(oxyethylene) block copolymers form reversible gels by the close packing of their micelles when concentrated solutions are warmed.

Swollen gels may exhibit *syneresis*, which is the separation of solvent phase from the gel. This may be explained as follows:

- During gel formation the polymer network is stretched as the gel swells by taking in the solvent.
- At equilibrium the contracting force of the polymer network is balanced by the swelling forces determined by the osmotic pressure.
- If the osmotic pressure decreases, for example on cooling or by changing the ionisation of the polymer molecules, water may be squeezed out of the gel and the gel appears to 'weep'.
- Syneresis may often be decreased by the addition of electrolyte, glucose and sucrose or by increasing the polymer concentration.

Polymers and macromolecules in solution may:

- *form complexes*, as for example between high-molecular-weight polyacids and polyethylene glycols, polyvinylpyrrolidone and poly(acrylic acid)s, and hyaluronic acid and the proteoglycans in the intracellular matrix in cartilage
- *bind ions* present in solution to form gels, as for example when Ca^{2+} ions are bound by alginate molecules
- *adsorb at interfaces*, for example:
- Insulin in solution will adsorb onto the inner surface of glass and poly(vinyl chloride) infusion containers and plastic tubing used in giving sets, so reducing its concentration in solution.
- Gelatin, acacia, poly(vinyl alcohol) will adsorb at the interface between oil and water in emulsions or on the surface of dispersed suspension particles and so stabilise these colloidal dispersions.

Tips

- Note that di- and triblock copolymers in which the A block is poly(oxyethylene) and the B block is poly(oxypropylene) are amphiphilic because poly(oxyethylene) is hydrophilic and poly(oxypropylene) is hydrophobic.
- These polymers are surface-active and may form micelles in aqueous solution.
- At high solution concentrations the micelles pack so closely that the solution becomes immobile, i.e. gelation occurs.
- Gelation may also occur when concentrated solutions are warmed because the solubility of poly(oxyethylene) decreases as temperature increases, i.e. it becomes more hydrophobic. Therefore more micelles form at the higher temperature and gelation occurs as they pack closely together.

- *imbibe large quantities of water.* This is utilised:
- in the manufacture of paper and sanitary towels, nappies and surgical dressings
- in the treatment of constipation and in appetite suppression.

Some water-soluble polymers used in pharmacy and medicine

Important examples

- Carboxypolymethylene (Carbomer, Carbopol):
- is a high-molecular-weight polymer of acrylic acid, containing a high proportion of carboxyl groups
- is used as a suspending agent in pharmaceutical preparations, as a binding agent in tablets, and in the formulation of prolonged-acting tablets.
- Cellulose derivatives:
- *Methylcellulose* is a methyl ether of cellulose containing about 29% of methoxyl groups. It is slowly soluble in water. Low-viscosity grades are used as emulsifiers for liquid paraffin and other mineral oils. High-viscosity grades are used as thickening agents for medicated jellies and as dispersing and thickening agents in suspensions.
- *Hydroxypropylmethylcellulose* (hypromellose) is a mixed ether of cellulose containing 27–30% of $-OCH_3$ groups and 4–7.5% of $-OC_3H_6OH$ groups. It forms a viscous colloidal solution and is used in ophthalmic solutions to prolong the action of medicated eye drops and is employed as an artificial tear fluid.
- Natural gums and mucilages:
- *Gum arabic* (acacia) is a very soluble polyelectrolyte whose solutions are highly viscous due to the branched structure of the macromolecular chains. It is used in pharmacy as an emulsifier.
- *Gum tragacanth* partially dissolves in water to give highly viscous solutions. It is one of the most widely used natural emulsifiers and thickeners and is an effective suspending agent.
- *Alginates* are block copolymer polysaccharides derived from seaweed consisting of β-D-mannuronic acid and

KeyPoints

- Water-soluble (hydrophilic) polymers are widely used in pharmacy, for example as suspending agents, emulsifiers, binding agents in tablets, thickeners of liquid dosage forms and in film coating of tablets.
- Water-insoluble (hydrophobic) polymers are mainly used in packaging material and tubing, and in the fabrication of membranes and films.
- Important properties of hydrophobic polymers which affect their suitability for use in pharmacy are their permeability to drugs and gases and their tendency to adsorb drugs.

α-L-guluronic acid residues joined by 1,4 glycosidic linkages. They form very viscous solutions and gel on addition of acid or calcium salts. They are used chiefly as stabilisers and thickening agents.

- *Pectin* is a purified carbohydrate product from extracts of the rind of citrus fruits and consists of partially methoxylated polygalacturonic acid. It readily gels in the presence of calcium or other polyvalent cations.

- *Chitosan* is a polymer obtained by the deacetylation of the polysaccharide chitin. The degree of deacetylation has a significant effect on the solubility and rheological properties of the polymer. Chitosan will form films, gels and matrices, making it useful for solid dosage forms, such as granules or microparticles.

- *Dextran* is a branched-chain polymer of anhydroglucose, linked through α-1,6 glucosidic linkages. Partially hydrolysed dextrans are used as plasma substitutes or 'expanders'.

- *Polyvinylpyrrolidone* is a homopolymer of *N*-vinylpyrrolidone used as a suspending and dispersing agent, a tablet-binding and granulating agent, and as a vehicle for drugs such as penicillin, cortisone, procaine and insulin to delay their absorption and prolong their action. It forms hard films which are utilised in film-coating processes.

- *Macrogols* (polyoxyethylene glycols) are liquid over the molecular weight range 200–700 and are used as solvents for drugs such as hydrocortisone. Higher-molecular-weight members of the series are semisolid and waxy and may be used as suppository bases.

Properties

Bioadhesivity

- Bioadhesivity arises from interactions between the polymer chains and the macromolecules on the mucosal surface – for maximum adhesion there should be maximum interaction (Figure 6.5).
- The charge on the molecules will be important – for two anionic polymers maximum interaction will occur when they are not charged.

Crystallinity

- Defects in the crystals allow preparation of microcrystals, e.g. microcrystalline cellulose (Avicel) by disruption of larger crystals.

Tip

Refer to Chapter 5 to see the mechanism of stabilisation of suspensions and emulsions by polymers.

Figure 6.5 Schematic representation of two phases, adhesive (A) and mucus (B), which adhere due to chain adsorption and consecutive chain entanglement during mucoadhesion. Reproduced Ffrom N.A. Peppas and A.G. Mikos. In *Bioadhesion* (R. Gurney and H. Junginger, eds). Wiss. Verlagsgesellschaft, Stuttgart, 1990.

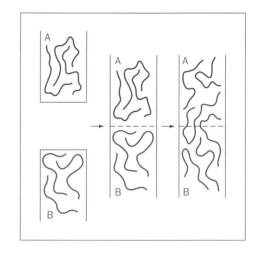

Water-insoluble polymers

Water-insoluble polymers play an important role in pharmacy and are used in the fabrication of membranes, containers and tubing. Important properties include:

- *Degree of crystallinity* – affects rigidity, fluidity, resistance to diffusion of small molecules and degradation.
- *Permeability to drugs*:
  - Diffusion of solutes in non-porous solid polymer is governed by Fick's first law, which for polymer membranes of thickness *l* becomes:

$$J = \frac{DK\Delta c}{l}$$

    where J is the flux, Δc represents the difference in solution concentration of drug at the two faces of membrane, D is the diffusion coefficient of the drug in the membrane and K is the distribution coefficient of the permeant towards the polymer.
  - Permeability within a given polymer is a function of the degree of crystallinity, which itself is a function of polymer molecular weight.
  - Permeation of drug molecules through the solid polymer is a function of the solubility of the drug in the polymer and will be altered by the presence of inorganic fillers in which drug is insoluble.
  - The permeability of a polymer film is affected by the method of preparation of the film.
  - Drug flux through dense (non-porous) polymer membranes is by diffusion; flux through porous membranes will be by diffusion and by transport in solvent through pores in the film.

- ▓ *Permeability to gases* – this is important when polymers are used as packaging materials:
- Permeability depends on the polarity of polymer – more polar films tend to be more ordered and less porous, hence less oxygen-permeable. The less polar films are more porous, permitting the permeation of oxygen but not necessarily of the larger water molecules.
- Water permeability can be controlled by altering the hydrophilic/hydrophobic balance of the polymer.
- ▓ *Affinity of drugs for plastics*:
- Steroids are adsorbed from solutions passing through polyethylene tubing.
- Glyceryl trinitrate has a high affinity for lipophilic plastics and migrates from tablets in contact with plastic liners in packages, causing a reduction of the active content of many tablets to zero.
- ▓ *Ion exchange properties*:
- Synthetic organic polymers comprising a hydrocarbon cross-linked network to which ionisable groups are attached have the ability to exchange ions attracted to their ionised groups with ions of the same charge present in solution. They are used in the form of beads as *ion exchange resins*.
- The resins may be either *cation* exchangers, in which the resin ionisable group is acidic, for example, sulfonic, carboxylic or phenolic groups, or *anion* exchangers, in which the ionisable group is basic, either amine or quaternary ammonium groups.
- *Cation exchange resins* effect changes in the electrolyte balance of the plasma by exchanging cations with those in the gut lumen. In the ammonium form, cation exchange resins are used in the treatment of retention oedema and for the control of sodium retention in pregnancy. These resins are also used (as calcium and sodium forms) to treat hyperkalaemia.
- *Anion exchange resins* such as polyamine methylene resin and polyaminostyrene have been used as antacids.
- Ion exchange resins are used in the removal of ionised impurities from water and in the prolongation of drug action by forming complexes with drugs.

Silicones are water-insoluble polymers with a structure containing alternate atoms of Si and O. Examples include:

- ▓ *Dimeticones* – fluid polymers with a wide range of viscosities. Commonly used as barrier substances, silicone lotions and creams acting as water-repellent applications protecting the skin against water-soluble irritants. Dimeticone 200 has been used as a lubricant for artificial eyes and to replace the degenerative vitreous fluid in cases of retinal detachment. It can also act as a simple lubricant in joints.

KeyPoint

Control of the rate of release of a drug when administered by oral or parenteral routes is aided by the use of polymers that function as a barrier to drug movement.

- *Activated dimeticone* (activated polymethylsiloxane) is a mixture of liquid dimeticones containing finely divided silica to enhance the defoaming properties of the silicone.

Application of polymers in drug delivery

Film coating

Polymer solutions allowed to evaporate produce polymeric films which can act as protective layers for tablets or granules containing sensitive drug substances or as a rate-controlling barrier to drug release.

- Film coats have been divided into two types: those that dissolve rapidly and those that behave as dialysis membranes allowing slow diffusion of solute or some delayed diffusion by acting as gel layers.
- Materials that have been used as film formers include shellac, zein, cellulose acetate phthalate, glyceryl stearates, paraffins, cellulose acetate phthalate, and a range of anionic and cationic polymers such as the Eudragit polymers.

Matrices

Methods of drug delivery from matrices include the use of:

- A non-eroding matrix: the mechanism of sustained release is:
  - the passage of drug through pores in the matrix if this is made of water-insoluble polymer (hydrophobic matrices)
  - the entry of water into the polymer matrix followed by swelling and gelation and then diffusion of drug through the viscous gel when water-soluble matrices (hydrophilic matrices) are used.
- A reservoir system – drug contained in the reservoir releases by leaching or slow diffusion through the wall of the retaining polymer membrane.
- An eroding matrix – drug is released when the polymer matrix in which a drug is dissolved or dispersed erodes by either bulk erosion or surface erosion (Figure 6.6).

In microcapsules drug is encapsulated as small particles or as a drug solution in a polymer film or coat, whereas microspheres are solid polymeric spheres which entrap drug. Equivalent structures with particle diameters ranging from 50 to 500 nm are referred to as *nanocapsules* and *nanoparticles*.

- *Microcapsules* can be prepared by three main processes:

1. *Coacervation* – the liquid or solid to be encapsulated is dispersed in a solution of a macromolecule (such as gelatin,

Figure 6.6 Matrix systems for drug delivery.

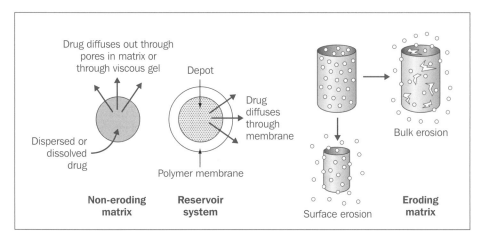

Drug diffuses out through pores in matrix or through viscous gel

Depot

Drug diffuses through membrane

Bulk erosion

Dispersed or dissolved drug

Polymer membrane

Non-eroding matrix

Reservoir system

Surface erosion

Eroding matrix

gum arabic, carboxymethylcellulose or poly(vinyl alcohol)) in which it is immiscible. A suitable non-solvent (such as ethanol and isopropranol) is added which is miscible with the continuous phase but a poor solvent for the polymer and causes the polymer to form a coacervate (polymer-rich) layer around the disperse phase. This coating layer may then be treated to give a rigid coat of capsule wall.

2. *Interfacial polymerisation* – reactions between oil-soluble monomers and water-soluble monomers at the oil–water interface of water-in-oil or oil-in-water dispersions can lead to interfacial polymerisation resulting in the formation of polymeric microcapsules, the size of which is determined by the size of the emulsion droplets. Alternatively, reactive monomer can be dispersed in one of the phases and induced to polymerise at the interface, or to polymerise in the bulk disperse phase and to precipitate at the interface due to its insolubility in the continuous phase.

3. *Physical methods.* The *spray drying* process involves dispersion of the core material in a solution of coating substance and spraying the mixture into an environment which causes the solvent to evaporate. The process of *pan coating* has been applied in the formation of sustained-release beads by application of waxes such as glyceryl monostearate in organic solution to granules of drug. It can only be used for particles greater than 600 μm in diameter.

▪ *Nanocapsules* may be prepared in a similar manner but the locus of polymerisation is not an emulsion droplet as in microencapsulation, but a micelle:

– The process involves the solubilisation of a water-soluble monomer such as acrylamide along with the drug or other agent such as antigen to be encapsulated.

- An organic liquid such as *n*-hexane serves as the outer phase.
- Polymerisation is induced by irradiation (γ-rays, X-rays, ultraviolet light), exposure to visible light or heating with an initiator.

■ *Microspheres and nanoparticles* can be prepared by modification of the coacervation process:
- For example, gelatin nanoparticles have been prepared by desolvation (for example, with sodium sulfate) of a gelatin solution containing drug bound to the gelatin, in a process which terminates the desolvation just before coacervation begins.
- In this manner colloidal particles rather than the larger coacervate droplets are obtained.
- Hardening of the gelatin nanoparticles is achieved by glutaraldehyde which cross-links with gelatin.

Rate-limiting membranes and devices

Utilise rate-limiting membranes to control the movement of drugs from a reservoir.

■ Drug release rate is controlled by choice of polymer, membrane thickness and porosity.
■ Examples include the Progestasert device designed to release progesterone into the uterine cavity, the Ocusert device for delivery to the eye and the Transiderm therapeutic system for transdermal medication.

Osmotic pumps

A variety of devices have been described:

■ In the *oral* osmotic pump (Oros) (Figure 6.7):
- The drug is mixed with a water-soluble core material.
- This core is surrounded by a water-insoluble semipermeable polymer membrane in which is drilled a small orifice.
- Water molecules can diffuse into the core through the outer membrane to form a concentrated solution inside.
- An osmotic gradient is set up across the semipermeable membrane with the result that drug is pushed out of the orifice.
- For example, the osmotic tablet of nifedipine consists of a semipermeable cellulose acetate coating, a swellable hydrogel layer of polyoxyethylene glycol and hydroxypropylmethylcellulose and a drug chamber containing nifedipine in hydroxypropylmethylcellulose and polyoxyethylene glycol.

■ There are two groups of *transdermal* delivery systems (Figure 6.8):
- *Membrane* systems generally consist of a reservoir, a rate-controlling membrane and an adhesive layer. Diffusion of the active principle from the reservoir through the controlling membrane governs release rate. The active principle is usually present in suspended form; liquids and gels are used as

dispersion media. Examples of membrane systems include Transiderm Nitro, in which the rate-controlling membrane is composed of a polyethylene/vinyl acetate copolymer having a thin adhesive layer and the reservoir contains nitroglycerin dispersed in the form of a lactose suspension in silicone oil.

– In *matrix* systems the active principle is dispersed in a matrix which consists of either a gel or an adhesive film. Examples of matrix systems include the Nitro-Dur system which consists of a hydrogel matrix (composed of water, glycerin, poly(vinyl alcohol) and polyvinylpyrrolidone) in which a nitroglycerin/lactose triturate is homogeneously dispersed.

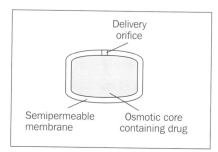

Figure 6.7 Diagrammatic representation of the Oros osmotic pump.

Figure 6.8 Structures of commercial transdermal types (a) membrane-controlled and (b) matrix systems.

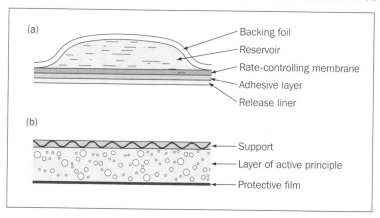

Multiple choice questions

In questions 1–7 indicate whether each of the statements is true or false.

1. **Polymers with regular alternation of the side chains on the polymer backbone are referred to as:**

a.   isotactic

b.   atactic

c.   syndiotactic

2. **The copolymer HO(CH$_2$CH$_2$O)$_x$ (CH(CH$_3$)CH$_2$O)$_y$(CH$_2$CH$_2$O)$_x$H is an example of a:**
a. graft copolymer
b. diblock copolymer
c. homopolymer
d. triblock copolymer
e. poloxamer

3. **The intrinsic viscosity of a polymer solution is:**
a. relative viscosity – 1
b. the ratio of the viscosity of the solution to the viscosity of pure solvent
c. proportional to the molecular weight of the polymer
d. obtained by extrapolation of plots of reduced viscosity against concentration to infinite dilution
e. of an approximate value of 2.5 for spherical, non-hydrated particles

4. **Cross-linked poly[HEMA] gels:**
a. are type II gels
b. can swell in water but cannot dissolve
c. are heat-reversible
d. may exhibit syneresis
e. may be used to form hydrophilic contact lenses

5. **The diffusion of small molecules through films formed from water-insoluble polymers:**
a. is not affected by the degree of crystallinity of the polymer
b. is a function of the polymer molecular weight
c. is not affected by the solubility of the molecule in the polymer
d. is governed by Fick's first law
e. is affected by the method of preparation of the polymer film

6. **Drug release from a non-eroding hydrophilic matrix drug release involves:**
a. swelling and gelation of the matrix
b. coacervation
c. diffusion of drug through a gel layer
d. diffusion of drug through a semipermeable membrane
e. passage of drug through pores in the matrix

7. **Typical microcapsules:**
a. may be prepared by interfacial polymerisation
b. release drugs by an osmotic gradient across a gel layer
c. have diameters of about 50 nm
d. may be prepared by a coacervation process
e. release drug through a small orifice

chapter 7
Drug absorption

Overview

In this chapter we will:

- review the structure and function of biological membranes and discuss the factors influencing the transport of drugs through them
- summarise the special features of a number of routes for drug administration either for systemic or local action, including:
  - Oral route and oral absorption
  - Buccal and sublingual absorption
  - Intramuscular (i.m.) and subcutaneous (s.c.) injection
  - Transdermal delivery
  - Eye and ear
  - Vaginal absorption
  - Lung and respiratory tract (inhalation therapy)
  - Nasal and rectal routes
  - Intrathecal drug administration.

Biological membranes and drug transport

Biological membranes

Membrane structure

- Figure 7.1 shows a diagram of the *fluid mosaic* model of a biological membrane.
- The fluid mosaic model, in particular, allows the protein–lipid complexes to form either hydrophilic or hydrophobic 'gates' to allow transport of materials of different characteristics.

Cholesterol

- Cholesterol is a major component of most mammalian biological membranes.
- Its shape allows it to fit closely in bilayers with the hydrocarbon chains of unsaturated fatty acids.
- Its removal causes the membrane to lose its structural integrity and to become highly permeable.
- It complexes with phospholipids and reduces the permeability of phospholipid membranes to water, cations, glycerol and glucose.

KeyPoints

- Absorption generally requires the passage of the drug in a molecular form across one or more barrier membranes and tissues.
- Most drugs are administered as solid or semisolid dosage forms.
- Tablets or capsules will disintegrate, and the drug will then be released for subsequent absorption.
- Many tablets contain granules or drug particles which should deaggregate to facilitate the solution process.
- If the drug has the appropriate physicochemical properties, it will pass by passive diffusion from a region of high concentration to a region of low concentration across the membrane.
- Soluble drugs can, of course, also be administered as solutions.

KeyPoints

- The main function of biological membranes is to contain the aqueous contents of cells and separate them from an aqueous exterior phase.
- Membranes are lipoidal in nature.
- To allow nutrients to pass into the cell and waste products to move out, biological membranes are *selectively permeable*.
- There are specialised transport systems to assist the passage of water-soluble materials and ions through their lipid interior.
- Some drugs are absorbed by transport systems and do not necessarily obey the pH partition hypothesis.
- Lipid-soluble agents can pass by passive diffusion through the membrane.
- Biological membranes are composed of bilayers of phospholipids and cholesterol or related structures.
- Embodied in the matrix of lipid molecules are proteins, generally hydrophobic in nature.
- Membranes have a hydrophilic, negatively charged exterior and a hydrophobic interior.

Drug absorption is controlled by the nature of the membrane, its degree of internal bonding and rigidity, its surface charge and by physicochemical properties of the drug. Note that cellular efflux mechanisms centred on P-glycoproteins exist. Some drugs are ejected from cells by these 'pumps' so that these drugs have a lower apparent absorption than predicted on physicochemical grounds.

Lipophilicity and absorption

If the percentage absorption versus log P is plotted a parabolic relationship is obtained, with the optimum value designated as log P_0 (Figure 7.2).

The optimal partition coefficient differs for different absorbing membranes.

Drugs with high log P values

- Drugs with high log P values are protein-bound, have low aqueous solubility and bind to extraneous sites.
- They have a lower bioavailability than anticipated from their log P, giving rise to the parabolic curve shown in Figure 7.2.

Drugs with low log P values

- May be too hydrophilic to have any affinity for the membranes and hence may be poorly absorbed.

Figure 7.1 Diagrammatic representation of the fluid mosaic model of a biological membrane.

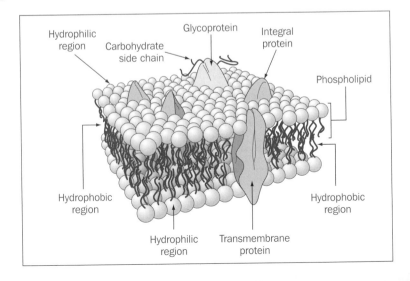

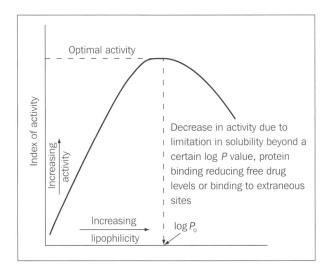

Figure 7.2 Typical activity–log P plot.

Molecular weight and drug absorption

■ The larger drug molecules are, the poorer will be their absorption.

■ Lipinski devised a *rule of five* defining drug-like properties. Good oral absorption is more likely when:

– the drug molecule has fewer than 5 hydrogen bond donors (–OH groups or –NH groups)

– the molecular weight of the drug is less than 500

– log P of the drug is less than 5

– there are fewer than 10 H-bond acceptors.

but note:

– Compounds that are substrates for transporters are exceptions to the rule.

Permeability and the pH-partition hypothesis

Assumption

■ Only drugs in their unionised form (more lipid-soluble) pass through membranes.

■ As most drugs are weak electrolytes it is to be expected that the unionised form (U) of either acids or bases, the lipid-soluble species, will diffuse across the membrane, while the ionised forms (I) will be rejected.

Calculation of percentage ionisation

■ For *weakly acidic drugs* (such as aspirin and indometacin) the ratio of ionised to unionised species is given by the equations:

$$pH - pK_a = \log\frac{[\text{ionised form}]}{[\text{unionised form}]} = \log\frac{[\text{I}]}{[\text{U}]}$$

Tips

The following example illustrates the calculation of the percentage ionisation from these equations.
The amount of drug in the unionised form of acetylsalicylic acid (pK_a = 3.5) at pH 4 is:

4 – 3.5 = log([I]/[U])
∴ [I]/[U] = 3.162
Percentage unionised
= ([U] × 100)/([U] + [I])
= 100/{1 + ([I]/[U])}
= 100/(1 + 3.162)
= (100/4.162) = 24.03%

- In choosing the correct equation you need to be able to distinguish between weakly acidic and basic drugs. You can usually tell the type of drug salt from the drug name. For example:
- The *sodium* of sodium salicylate means that it is the salt of the strong base *sodium* hydroxide and the weak acid salicylic acid, i.e. sodium (or potassium) in the drug name implies the salt of a strong base.
- The *hydrochloride* of ephedrine hydrochloride means that it is the salt of a strong acid (*hydrochloric* acid) and a weak base (ephedrine), i.e. hydrochloride (or bromide, nitrate, sulfate, etc.) in the drug name implies the salt of a strong acid.

- For *weakly basic* drugs the equation takes the form:

$$pH - pK_a = \log \frac{[\text{unionised form}]}{[\text{ionised form}]} = \log \frac{[U]}{[I]}$$

Discrepancies between expected and observed absorption

- Absorption is often much greater than one would expect, although the trend is as predicted. For example, absorption of acetylsalicylic acid is 41% at pH 4 (although only 24% is in unionised form) and 27% at pH 5 (only 3.1% is in unionised form).
- Two possible explanations:
  1. Absorption and ionisation are both dynamic processes so that even small amounts of unionised drug can be absorbed and are replenished.
  2. The bulk pH is not the actual pH at the membrane. A local pH exists at the membrane surface which differs from the bulk pH. This local pH is lower than the bulk because of the attraction of hydrogen ions by the negative groups of the membrane components (see below).

Problems in the quantitative application of the pH-partition hypothesis

There are several reasons why the pH-partition hypothesis cannot be applied quantitatively in practical situations:

Variability in pH conditions

- Although the normally quoted range of stomach pH is 1–3, studies using pH-sensitive radiotelemetric capsules have shown a greater spread of values, ranging up to pH 7.
- The scope for variation in the small intestine is less, although in some pathological states the pH of the duodenum may be quite low due to hypersecretion of acid in the stomach.

pH at membrane surfaces

- pH at the membrane surface is lower than that of the bulk pH, hence the appropriate pH has to be inserted into equations, and the solubility of drug will change in the vicinity of the membrane.
- The secretion of acidic and basic substances in many parts of the gut wall is also a complicating factor.

Convective water flow

- The movement of water molecules, due to differences in osmotic pressure between blood and the contents of the lumen and differences in hydrostatic pressure between lumen and perivascular tissue, affects the rate of absorption of small molecules.
- Absorption of water-soluble drugs will be increased if water flows from the lumen to the serosal blood side across the mucosa, provided that drug and water are using the same route.
- Water movement is greatest within the jejunum.

Unstirred water layers

- A layer of relatively unstirred water lies adjacent to all biological membranes.
- During absorption drug molecules must diffuse across this layer, which is an additional barrier.

Effect of the drug

The drug must be in its molecular form before diffusional absorption processes take place.

- Basic drugs are therefore expected to be more soluble than acidic drugs in the stomach.
- Although the basic form of a drug as its hydrochloride salt should be soluble to some extent in this medium, this is not always so. The free bases of chlortetracycline and methacycline are more soluble than their hydrochloride salts in the stomach.

Other complicating factors

- The very high area of the surface of the small intestine also upsets the calculation of absorption based on considerations of theoretical absorption across identical areas of absorbing surface.
- Co-administration of drugs such as cimetidine can raise stomach pH from below 2 to near neutrality.
- The following may not be absorbed as expected: drugs which are:
- unstable in the gastrointestinal tract (for example, erythromycin)

KeyPoints

- The nature of the formulation often has a large effect on drug absorption from some sites.
- The important features are always the interplay between:
- drug
- vehicle (formulation)
- the route of administration.
- The same drug may be absorbed from different sites, often in quite different amounts. For example: cocaine with a *P* of 28 requires a sublingual/s.c. dosage ratio of 2 to obtain equal effects, atropine with a *P* of 7 requires eight times the s.c. dose, and for codeine (*P* ~ 2) over 15 times the s.c. dose must be given sublingually.

KeyPoints

- The functions of the gastrointestinal tract are the digestion and absorption of foods and nutrients and it is not easy to separate these from drug delivery.
- The natural processes in the gut can influence the absorption of drugs.
- The pH of the gut contents and the presence of enzymes, foods, bile salts, fat and the microbial flora can influence drug absorption.
- The complexity of the absorbing surfaces means that a simple physicochemical approach to drug absorption remains an *approach* to the problem and not the complete picture, as described above.

- metabolised on their passage through the gut wall
- hydrolysed in the stomach to active forms (prodrugs)
- bound to mucin to form complexes with bile salts
- in the charged form, which interacts with other ions to form absorbable species with a high lipid solubility – ion pair formation.

Routes of administration

The oral route and oral absorption

Drug absorption from the gastrointestinal tract

Factors affecting absorption from oral dosage forms (in addition to the properties of the drug) include:

- the extent and rate of dissolution of the drug
- the rate of gastric emptying
- the site of absorption.

Structure of the gastrointestinal tract (Figure 7.3)

The stomach

- The stomach is not an organ designed for absorption.
- Its volume varies with the content of food (it may contain a few millilitres or a litre or more of fluid).
- Hydrochloric acid is liberated from the parietal cells (at a concentration of 0.58%).
- The gastric glands produce around 1000–1500 cm^3 of gastric juice per day.

The small intestine

- The small intestine is the main site of absorption.
- The small intestine is divided anatomically into three sections, *duodenum*, *jejunum* and *ileum*, with no clear transition between them.

- All three are involved in the digestion and absorption of foodstuffs; absorbed material is removed by the blood and the lymph.
- The absorbing area is enlarged by surface folds in the intestinal lining which are macroscopically apparent: the surface of these folds possesses villi (Figure 7.4).
- The human small intestine has a calculated active surface area of approximately 100 m².
- The surface area of the small intestine of the rat is estimated to be 700 cm², a difference of 1440-fold.
- Differences in the absorptive areas and volumes of gut contents in different animals are important when comparing experimental results on drug absorption in various species.

Figure 7.3 Representation of the processes occurring along the gastrointestinal tract.

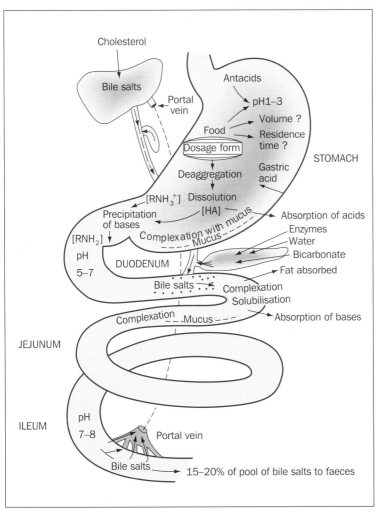

Figure 7.4 Representation of the epithelium of the small intestine showing intestinal villi and microvilli.

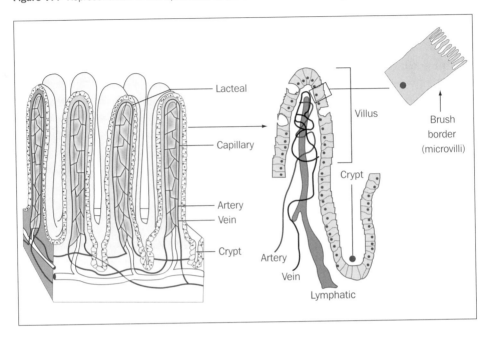

The large intestine

- The large intestine is primarily concerned with the absorption of water and the secretion of mucus to aid the intestinal contents to slide down the intestinal 'tube'.
- Villi are completely absent from the large intestine.

Carrier-mediated and specialised transport

There is the possibility of absorption of drugs by way of:

- tight junctions (the paracellular route)
- carrier-mediated uptake mechanisms
- endocytosis (Figure 7.5).

Figure 7.5 Gastrointestinal membrane transport.

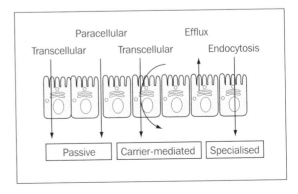

Bile salts and fat absorption pathways

- Fat is absorbed by special mechanisms in the gut.
- Bile salts secreted into the jejunum are efficient emulsifiers and disperse fat globules, so increasing the surface area for absorption.
- Lipase activity is enhanced at the resulting large surface area.
- Medium-chain triglycerides are thought to be directly absorbed.
- Long-chain triglycerides are hydrolysed.
- Monoglycerides and fatty acids produced form mixed micelles with the bile salts and are either absorbed directly in the micelle or, more probably, brought to the microvillous surface by the micelle and transferred directly to the mucosal cells.
- There have been suggestions that lipid-soluble drugs may be absorbed by fat absorption pathways.
- The administration of drugs in an oily vehicle can significantly affect absorption, e.g. of griseofulvin and ciclosporin.

Tips

Remember that:
- Bile salts are amphiphilic compounds and can form micelles or aggregates in solution in the gastrointestinal tract.
- Monoglycerides and fatty acids are also amphiphilic and so will be incorporated into the bile salt micelles, resulting in mixed micelle formation.

Gastric emptying, motility and volume of contents

- The volume of the gastric contents will determine the concentration of a drug which finds itself in the stomach.
- The time the drug or dosage form resides in the stomach will determine many aspects of absorption from solid dose forms:
  - If the drug is absorbed in the intestine, emptying rates will determine the delay before absorption begins.
  - If a drug is labile in acid conditions, longer residence times in the stomach will lead to greater breakdown.
  - If the dosage form is non-disintegrating then retention in the stomach can influence the pattern of absorption.
- The stomach empties liquids faster than solids.
- Gastric emptying is a simple exponential or square-root function of the volume of a test meal – a pattern that holds for meals of variable viscosity.
- Acids have been found to slow gastric emptying, but acids with relatively high molecular weights (for example, citric acid) are less effective than those, such as hydrochloric acid, with very low molecular weights.
- When considering the effect of an antacid, therefore, the effect of volume change and pH change and the effect on gastric emptying must be considered.
- Food affects not only transit but also pH in the gastrointestinal tract.
- Natural triglycerides inhibit gastric motility.

KeyPoint

The formulation of a drug may influence drug absorption through an indirect physiological effect, e.g. whether it is solid or liquid, acidic or alkaline, aqueous or oily, may influence gastric emptying.

Buccal and sublingual absorption

- The absorption of drugs through the oral mucosa provides a route for systemic administration.
- This route avoids exposure of drug to the gastrointestinal tract.
- Drugs bypass the liver (and so avoid metabolism there) and have direct access to the systemic circulation.
- Drugs such as glyceryl trinitrate have traditionally been administered in this way. This drug exerts its pharmacological action 1–2 minutes after sublingual administration.
- Mucosal adhesive systems are used for administration of buprenorphine through the gingiva (gums).

Mechanisms of absorption

- The oral mucosa functions primarily as a barrier and it is not highly permeable. It comprises:
- a mucous layer over the epithelium
- a keratinised layer in certain regions of the oral cavity
- an epithelial layer
- a basement membrane
- connective tissue
- a submucosal region.
- Most drugs are absorbed by simple diffusion.
- There is a linear relationship between percentage absorption through the buccal epithelium and log P of a homologous drug series.
- Buccal absorption of basic drugs increases and that of acidic drugs decreases, with increasing pH of their solutions.
- Nicotine in a gum vehicle is absorbed through the buccal mucosa.
- The buccal route has the advantages of the sublingual route – the buccal mucosa is similar to sublingual mucosal tissue – but a sustained-release tablet can be held in the cheek pouch for several hours if necessary.
- A log P of 1.6–3.3 is optimal for drugs to be used by the sublingual route.

Intramuscular and subcutaneous injection

- Not all drugs are efficiently or uniformly released from i.m. or s.c. sites.
- The s.c. region has a good supply of capillaries, although there are few lymph vessels in muscle proper.
- Drugs can diffuse through the tissue and pass across the capillary walls and thus enter the circulation via the capillary supply.

- Molecular size of soluble drugs is important: mannitol rapidly diffuses from the site of injection, insulin (molecular weight ~6000) less rapidly and dextran (molecular weight 70 000) disperses more slowly.
- Hydrophobic drugs may bind to muscle protein, leading to a reduction in free drug and perhaps to prolongation of action:
- Dicloxacillin is 95% bound to protein; ampicillin is bound to the extent of 20%, and as a consequence dicloxacillin is absorbed more slowly from muscle than is ampicillin.

Site of injection

- The regions into which the injection is administered are composed of both aqueous and lipid components (Figure 7.6).
- Muscle tissue is more acidic than normal physiological fluids.
- The pH of the tissue will determine whether or not drugs will dissolve in the tissue fluids or precipitate from formulations.
- The deliberate reduction of the solubility of a drug achieves prolonged action by both routes.

Figure 7.6 Routes of parenteral medication. Modified from Quackenbusch D S (1969) In: Martin E W (ed.) *Techniques of Medication*. Philadelphia, PA: Lippincott.

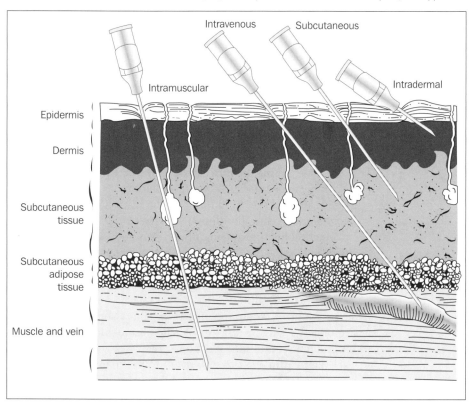

Vehicles

- Many i.m. injections are formulated in water-miscible solvents such as polyethylene glycol 300 or 400, or propylene glycol or ethanol mixtures.
- Dilution by the tissue fluids may cause a drug to precipitate.
- Three main types of formulation are used:
  1. aqueous solutions, which are rapidly removed
  2. aqueous suspensions
  3. oily solutions from which drugs (e.g. fluphenazine decanoate) can diffuse slowly for long action.

Blood flow

- Different rates of blood flow in different muscles mean that the site of i.m. injection can be crucial.
- Resting blood flow in the deltoid region is significantly greater than in the *gluteus maximus* muscle; flow in the *vastus lateralis* is intermediate.

Formulation effects

- Crystalline suspensions of fluspirilene, certain steroids and procaine benzylpenicillin can be prepared in different size ranges to produce different pharmacokinetic profiles following i.m. or s.c. injection.
- Variability in response to a drug, or differences in response to a formulation from different manufacturers, can be the result of the nature of the formulation.
- The depth of the injection is significant. If, in addition, the blood supply to the region is limited there will be an additional restriction to rapid removal.

Insulin

- A classic example of what can be achieved by manipulation of the properties of the drug and formulation.
- Modification of the crystallinity of the insulin allows control over solubility and duration of activity.
- Long-acting insulins are mainly protamine insulin and zinc insulins.
- Protamine insulins are salt-like compounds formed between the acid (insulin) polypeptide and the protamine polypeptide (primarily of arginine). They are used in the form of neutral suspensions of protamine insulin crystals (isophane insulin).
- Prolonged-acting insulins have been designed to have intermediate durations of action.
- Variable insulin activity may result from the mixing of protamine-zinc-insulin and soluble insulin prior to administration.

- As a general rule, insulin formulations of different pH should not be mixed.

Transdermal delivery

The barrier layer of the skin is the *stratum corneum*, which behaves like a passive diffusion barrier (Figure 7.7).

Figure 7.7 Simplified model of the *stratum corneum*, illustrating possible pathways of drug permeation. Reproduced from Moghimi H R et al., *Int. J. Pharm* 1996; 131.117.

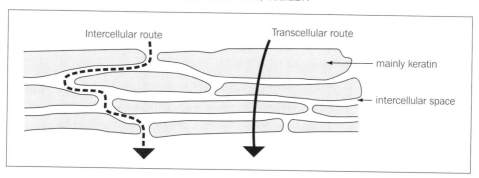

The vehicle in which the drug is applied influences the rate and extent of absorption, but formulations can change rapidly once they have been spread on the skin, with absorption of some excipients and evaporative loss of water.

Routes of skin penetration

- Solute molecules may penetrate the skin not only through the *stratum corneum* but also by way of the hair follicles or through the sweat ducts.
- Only in the case of molecules that move very slowly through the *stratum corneum* may absorption by these other routes predominate.
- The major pathway of transport for water-soluble molecules is *transcellular*, involving passage through cells and cell walls.
- The pathway for lipid-soluble molecules is presumably the endogenous lipid within the *stratum corneum*; the bulk of this is *intercellular*.
- Passage through damaged skin is increased over normal skin. For example, skin with a disrupted epidermal layer will allow up to 80% of hydrocortisone to pass into the dermis but only 1% through intact skin.

Tip

Occlusive dressings are those that have low permeability to water vapour. When these are left on the skin they prevent water loss from the skin surface and therefore increase the hydration of the *stratum corneum*.

- The physicochemical factors that control drug penetration include:
- the hydration of the *stratum corneum* (occluded skin may absorb up to 5–6 times its dry weight of water – see Tip)
- temperature
- pH
- drug concentration
- the molecular characteristics of the penetrant
- the vehicle.

Influence of drug

- The diffusion coefficient of the drug in the skin will be determined by factors such as molecular size, shape and charge.
- The effective partition coefficient will be determined not only by the properties of the drug but also by the vehicle as this represents the donor phase, the skin being the receptor phase.
- The peak activity in a drug series coincides with an optimal partition coefficient – one that favours neither lipid nor aqueous phase. Examples include:
- Triamcinolone is 5 times more active systemically than hydrocortisone, but has only one-tenth of its topical activity.
- Triamcinolone acetonide has a topical activity 1000 times that of the parent steroid because of its favourable lipid solubility.
- Betamethasone is 30 times as active as hydrocortisone systemically but has only four times the topical potency.
- Of 23 esters of betamethasone, the 17-valerate ester possesses the highest topical activity. The vasoconstrictor potency of betamethasone 17-valerate is 360 (fluocinolone acetonide = 100), its 17-acetate 114, the 17-propionate 190, and the 17-octanoate, 10.
- Not all drugs are suitable for transdermal delivery.

Formulations

- Many are washable oil-in-water systems.
- Simple aqueous lotions are also used as they have a cooling effect on the skin.
- Ointments are used for the application of insoluble or oil-soluble medicaments and leave a greasy film on the skin, inhibiting loss of moisture and encouraging the hydration of the keratin layer.
- Aqueous creams combine the characteristics of the lotions and ointments.
- Ointments are generally composed of single-phase hydrophobic bases, such as pharmaceutical grades of soft paraffin or microcrystalline paraffin wax.

- 'Absorption' bases have a capacity to facilitate absorption by the skin but the term also alludes to their ability to take up considerable amounts of water to form water-in-oil emulsions.

Drug release from vehicles

- In emulsions the relative affinity of drug for the external and internal phases is an important factor.
- A drug dissolved in an internal aqueous phase of a water-in-oil emulsion must be able to diffuse through the oily layer.
- Three cases can be considered: (1) solutions; (2) suspensions; and (3) emulsion systems.

Solutions

- Release rate is proportional to the square root of the diffusion coefficient and hence release is slower from a viscous vehicle.

Suspensions

- Release rate is proportional to the square root of the total solubility of the drug in the vehicle and hence for a drug in suspension to have any action it must have a degree of solubility in the base used.

Emulsion systems

- Release rate is proportional to the diffusion coefficient of the drug in the continuous phase and inversely proportional to partition coefficient between the phases.

Patches and devices

The ease with which some drugs can pass through the skin barrier into the circulating blood means that the transdermal route of medication is a possible alternative to the oral route. Theoretically there are several advantages:

- For drugs that are normally taken orally, administration through the skin can eliminate the vagaries that influence gastrointestinal absorption, such as pH changes and variations in food intake and intestinal transit time.
- A drug may be introduced into the systemic circulation without initially entering the portal circulation and passing through the liver.
- Constant and continuous administration of drugs may be achieved by a simple application to the skin surface.
- Continuous administration of drugs percutaneously at a controlled rate should permit elimination of pulse entry into the systemic circulation, an effect that is often associated with side-effects.

■ Absorption of medication could be rapidly terminated whenever therapy must be interrupted.

Patches

There are currently four basic forms of patch systems (Figure 7.8):

1. matrix
2. reservoir
3. multilaminate
4. drug-in-adhesive.

Figure 7.8 The four main types of transdermal patch. Courtesy of 3M.

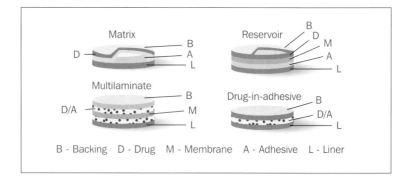

Iontophoresis

■ Iontophoresis is the process by which the migration of ionic drugs into tissues is enhanced by the use of an electrical current.

■ Enhancement of permeability results from several possible sources:

– ion–electric field interaction (electrorepulsion)
– convective flow (electro-osmosis)
– current-induced increases in skin permeability.

Ultrasound and transdermal penetration

■ Therapeutic ultrasound first expands and then collapses air bubbles in the *stratum corneum* (the process of *cavitation*).

■ Cavitation tends to liquefy the solid fats and allows molecules such as insulin to pass through the skin.

■ The permeability of the skin increases as the frequency of ultrasound decreases.

Jet injectors

■ Systems are based on the high-velocity ejection of particles through an orifice.

- Drug delivery is then possibly due to either or both of skin 'failure' and possibly convective flow through the skin.

The eye

- A wide range of drug types are placed in the eye, including antimicrobials, antihistamines, decongestants, mydriatics, miotics and cycloplegic agents.
- Drugs are usually applied to the eye in the form of drops or ointments for local action.
- The absorbing surface is the cornea.
- Drug absorbed by the conjuctiva enters the systemic circulation.
- The eye has two barrier systems:
  1. a blood–aqueous barrier.
  2. a blood–vitreous barrier.

Tears

- Tears contain electrolytes – sodium, potassium and some calcium ions, chloride and other counterions and glucose.
- Macromolecular components include some albumin, globulins and lysozyme.
- Lipids form a monolayer over the tear fluid surface.
- Drugs or excipients may interact with components of the tear fluid, so that tear coverage of the eye is disrupted.
- Dry-eye syndrome (xerophthalmia) may arise because of premature break-up of the tear layer, resulting in dry spots on the corneal surface.

Absorption of drugs applied to the eye

The *cornea* is the main barrier to absorption and comprises an epithelium, a stroma and an endothelium:

- The endothelium and the epithelium have a high lipid content and are penetrated by drugs in their unionised lipid-soluble forms.
- The stroma lying between the two other structures has a high water content. Thus drugs which have to negotiate the corneal barrier successfully must be both lipid-soluble and water-soluble to some extent.

Aqueous humour

- Both water-soluble and lipid-soluble drugs can enter the aqueous humour.
- The pH-partition hypothesis accounts only imperfectly for different rates of entry into aqueous humour.

KeyPoints

- As tears have some buffering capacity, the pH-partition hypothesis for drug absorption has to be applied with some circumspection.
- However, in agreement with the pH-partition hypothesis, raising the pH from 5 to 8 results in a two- to threefold increase in the amount of pilocarpine reaching the anterior chamber.

Influence of formulation

Some ingredients of eye medications may increase the permeability of the cornea.

Surface-active agents are known to interact with membranes to increase permeability:

- Benzalkonium chloride (bacteriostat and bactericide) has surfactant properties and may well have some effect on corneal permeability.
- Chlorhexidine acetate and cetrimide, both of which are surface-active, are also used.

Eye drops

- Are usually formulated to be isotonic with tear fluid.
- The rate of drainage of drops decreases as their viscosity increases and this can contribute to an increased concentration of the drug in the precorneal film and aqueous humour:
  - Hydrophilic polymeric vehicles, such as poly(vinyl alcohol) and hydroxypropylmethylcellulose are used to adjust viscosity.
- Most of the dose applied to the eye in the form of drops reaches the systemic circulation and typically less than 5% acts on ocular tissues.

Prodrugs

A prodrug of adrenaline (epinephrine), the dipivoyl derivative of adrenaline, is absorbed to a greater extent and is then hydrolysed to the active parent molecule in the aqueous humour.

Reservoir systems

- Soft lenses can be used as drug reservoirs leaching drug over 24 h.
- The Alza Ocusert device releases controlled amounts of pilocarpine over a period of 7 days.

The ear

- Medications are administered to the ear only for local treatment.
- Drops and other vehicles administered to the ear will occupy the external auditory meatus, which is separated from the middle ear by the tympanic membrane.
- The acidic environment of the ear skin surface (around pH 6), sometimes referred to as the acid mantle of the ear, is thought to be a defence against invading microorganisms.

Absorption from the vagina

- The vagina cannot be considered to be a route for the systemic administration of drugs, although oestrogens for systemic delivery have been applied intravaginally.
- Certain drugs, however, are absorbed when applied to the vaginal epithelium. Steroids, prostaglandins, iodine and some antibiotics and antifungals such as econazole and miconazole are appreciably absorbed.
- The pH in the vagina decreases after puberty, varying between pH 4 and 5 depending on the point in the menstrual cycle and also on the location within the vagina, the pH being higher near the cervix.
- There is little fluid in the vagina.
- The absorbing surface is under constant change, therefore absorption is variable.
- Mucus may retard absorption.
- Lymph vessels drain the vagina, and vaginal capillaries are found in close proximity to the basal epithelial layer.

Formulations

- Conventional vaginal delivery systems include vaginal tablets, foams, gels, suspensions and pessaries.
- Vaginal rings have been developed to deliver contraceptive steroids. These commonly comprise an inert silicone elastomer ring covered with an elastomer layer containing the drug.
- Hydrogel-based vaginal pessaries to deliver prostaglandin E_2 and bleomycin have been developed.
- Tablets often contain excipients which increase viscosity and are bioadhesive, e.g. hydroxypropyl cellulose, sodium carboxymethylcellulose and poly(acrylic acid) (such as Carbopol 934).
- Micropatches in the size range 10–100 μm in diameter prepared from starch, gelatin, albumin, collagen or dextrose will gel on contact with vaginal mucosal surfaces and adhere.

Inhalation therapy

- The respiratory system provides a route of medication.
- The contact area of its surfaces extends to more than 30 m^2.
- There are 2000 km of capillaries in the lungs.
- The route has been widely used in attempts to avoid systemic side-effects, such as adrenal suppression, but evidence suggests that inhaled steroids are absorbed systemically to a significant extent.

■ The respiratory tract epithelium has permeability characteristics similar to those of the classical biological membrane, so lipid-soluble compounds are absorbed more rapidly than lipid-insoluble molecules.

■ Compared to the gastrointestinal mucosa, the pulmonary epithelium possesses a relatively high permeability to water-soluble molecules, which is an advantage with drugs such as sodium cromoglicate.

■ The efficiency of inhalation therapy is often low because of the difficulty in targeting particles to the sites of maximal absorption.

■ Only about 8% of the inhaled dose of sodium cromoglicate administered from a dry-powder device reaches the alveoli.

Physical factors affecting deposition of aerosols

■ Deposition of particles in the various regions of the respiratory tract is dependent on particle size (Figure 7.9).

■ The major processes that influence deposition of drug particles in the respiratory tract are:
- interception
- impaction
- gravitational settling
- electrostatic attractions
- Brownian diffusion.

■ Very fine particles (< 0.5 µm) are deposited on the walls of the smallest airways by diffusion, the result of bombardment of the particles by gas molecules.

■ Particle size, or particle size distribution, is obviously important in several of these processes and will be affected by the nature of the aerosol-producing device and by the formulation.

Particles of hygroscopic materials are removed from the air stream more effectively than are non-hygroscopic particles, because of their growth through uptake of water from the moist air in the respiratory tract.

Delivery devices

Pressurised aerosols

■ Single-phase and two-phase systems are utilised.

■ In two-phase systems the propellant forms a separate liquid phase, whereas in the single-phase form the liquid propellant is the liquid phase containing the drug in solution or in suspension in the liquified propellant gas.

Tips

Difference between physical diameter and aerodynamic diameter

■ The aerodynamic diameter of a particle, d_a, is related to the particle diameter (d) and density (ρ) by the equation:

$$d_a = \rho^{0.5}d$$

■ To overcome problems of powder flow and agglomeration, porous particles (i.e. particles with a low density) have been developed. A particle of 10 µm diameter with a density of 0.1 g cm^{-3} has an aerodynamic diameter ~3 µm (i.e. $(0.1)^{0.5} \times 10$).

Figure 7.9 Deposition of particles in the various regions of the respiratory tract.

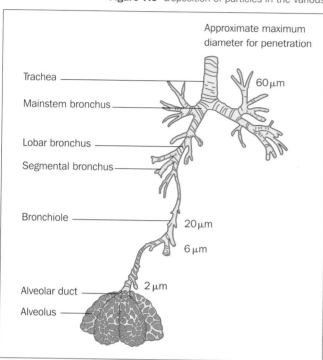

Approximate maximum diameter for penetration

Trachea — 60 μm

Mainstem bronchus

Lobar bronchus

Segmental bronchus

Bronchiole — 20 μm

6 μm

Alveolar duct — 2 μm

Alveolus

Nebulisers

- Modern nebulisers for domestic and hospital use generate aerosols continuously for chronic therapy of respiratory disorders.
- The particle size distribution varies with the design and sometimes mode of use.

The nasal route

Three main classes of medicinal agents are applied by the nasal route:

1. drugs for the alleviation of nasal symptoms
2. drugs that are inactivated in the gastrointestinal tract following oral administration
3. where the route is an alternative to injection, such as peptides and proteins.

- Delivery of peptides and proteins such as insulin, luteinizing hormone-releasing hormone analogues such as nafarelin, vasopressin, thyrotropin-releasing hormone analogues and adrenocorticotrophic hormone is feasible.
- Factors such as droplet or particle size which affect deposition in the respiratory tract are involved if administration is by aerosol (Figure 7.10).

■ The physiological condition of the nose, its vascularity and mucus flow rate are therefore of importance.

■ Formulation factors include:

– the volume

– concentration

– viscosity

– pH

– tonicity of the applied medicament.

■ As with all routes, absorption decreases with the increasing molecular weight of the active.

Figure 7.10 Regional depositional of inhaled particles as a function of aerodynamic diameter. Reproduced from C.D.F. Muir. *Clinical Aspects of Inhaled Particles,* Heinemann, London, 1972.

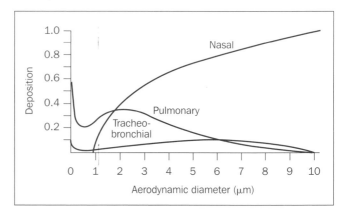

Rectal absorption of drugs

■ Drugs administered by the rectal route in *suppositories* are placed in intimate contact with the rectal mucosa, which behaves as a normal lipoidal barrier.

■ The pH in the rectal cavity lies between 7.2 and 7.4, but the rectal fluids have little buffering capacity.

■ As with topical medication, the formulation of the suppository can have marked effects on the activity of the drug.

■ Factors such as *retention* of the suppository for a sufficient duration of time in the rectal cavity also influence the outcome of therapy; the *size* and *shape* of the suppository and its *melting point* may also determine bioavailability.

■ Modern suppository vehicles include polyoxyethylene glycols of molecular weight 1000–6000 and semisynthetic vegetable fats.

■ The appropriate bases must be carefully selected for each substance. The important features of excipient materials are melting point, speed of crystallisation and emulsifying capacity.

- If the medicament dissolves in the base it is likely that the melting point of the base will be lowered, so that a base with a melting point higher than 36–37°C has to be chosen.
- If the drug substance has a high density it is preferable that the base crystallises rapidly during production of the suppositories to prevent settling of the drug.

The rectal cavity

- The rectum is the terminal 15–19 cm of the large intestine.
- The mucous membrane of the rectal ampulla, with which suppositories come into contact, is made up of a layer of cylindrical epithelial cells without villi.
- The main artery to the rectum is the superior rectal (haemorrhoidal) artery.
- Veins of the inferior part of the submucous plexus become the rectal veins, which drain to the internal pudendal veins.
- Drug absorption takes place through this venous network. Superior haemorrhoidal veins connect with the portal vein and thus transport drugs absorbed in the upper part of the rectal cavity to the liver; the inferior veins enter into the *inferior vena cava* and thus bypass the liver.

Fate of drug

- The particular venous route the drug takes is affected by the extent to which the suppository migrates in its original or molten form further up the gastrointestinal tract, and this may be variable.
- The rectal route does not necessarily, or even reproducibly, avoid the liver.

Absorption from formulations

- The rate-limiting step in drug absorption for suppositories made from a fatty base is the partitioning of the dissolved drug from the molten base, not the rate of solution of the drug in the body fluids.
- Absorption from the rectum depends on the concentration of drug in absorbable form in the rectal cavity and, if the base is not emulsified, on the contact area between molten excipient and rectal mucosa.
- Water-soluble active substances will be insoluble in fatty bases, while the less water-soluble material will tend to be soluble in the base, and will thus diffuse from the base more slowly.
- Water-soluble drugs are better absorbed from a fatty excipient than from a water-soluble one.
- Addition of surfactants may increase the ability of the molten mass to spread and tends to increase the extent of absorption.

- Hygroscopicity of some hydrophilic bases such as the polyoxyethylene glycols results in the abstraction of water from the rectal mucosa, causing stinging and discomfort, and probably affects the passage of drugs across the rectal mucosa.

Incompatibility between base and drug

Various incompatibilities exist:

- Phenolic substances complex with glycols, probably by hydrogen bonding between the phenolic hydroxy group and the glycol ether oxygens.
- Polyoxyethylene glycol bases are incompatible with tannic acid, ichthammol, aspirin, benzocaine, vioform and sulfonamides.
- Glycerogelatin bases are prepared by heating together glycerin, gelatin and water:
  - Use of untreated gelatin renders the base incompatible with acidic and basic drugs.
  - Two types of treated gelatin are employed with different characteristics to avoid incompatibilities:
    - Type A is acidic and cationic, with an isoelectric point between pH 7 and 9.
    - Type B is less acidic and anionic, with an isoelectric point between pH 4.7 and 5.

Intrathecal drug administration

- Administration of drugs in solution by intrathecal catheter provides an opportunity to deliver drugs to the brain and spinal cord (Figure 7.11).
- Relatively hydrophilic drugs such as methotrexate ($\log P = -0.5$), which do not cross the blood–brain barrier in significant amounts, have been infused intrathecally to treat meningeal leukaemia, and baclofen ($\log P = -1.0$) to treat spinal cord spasticity.
- High lumbar cerebrospinal fluid concentrations are achieved.
- The spinal cerebrospinal fluid has a small volume (70 cm³) and a relatively slow clearance (20–40 cm³ h⁻¹) for hydrophilic drugs.

KeyPoints

In summary:

- Each route has its own special characteristics. The nature of the absorption barrier in each is discussed on the basis of differences in liquid volume, pH, blood flow and drainage.
- The optimal lipophilicity of absorbing membranes depends on the nature of the membrane.
- Some barriers (as in the eye and skin) are complex, having the characteristics of typical lipid barriers, interspersed with more aqueous hurdles.
- In some cases (e.g. i.m. injections) the nature of the surrounding tissue, whether fatty or aqueous, is the key to the process of transferring drug into the blood.
- The overriding importance of lipophilicity is clear when drug is absorbed in molecular form, although water-soluble drugs can gain access, e.g. in the lung due to the very large surface area of contact between the absorbing membrane and the blood.
- When drug is delivered as a suspension (as in an aerosol) the paramount importance of particle size in first getting the drug to the site of action is clear; once it has reached that site (the alveoli), its rate of solution and its lipophilicity are again important.

- The cerebrospinal fluid pharmacokinetics of three drugs, morphine (log P = 0.15.), clonidine (log P = 0.85) and baclofen (log P = −1.0), were found to be similar, leading to the suggestion that bulk flow mechanisms may be the dominant factor in determining distribution.

Figure 7.11 Anatomical structures and absorption routes for a drug administered via an intrathecal catheter. CSF, cerebrospinal fluid. Reproduced From Kroin J S, *Clin. Pharmacokinet.* 22, 319 (1992).

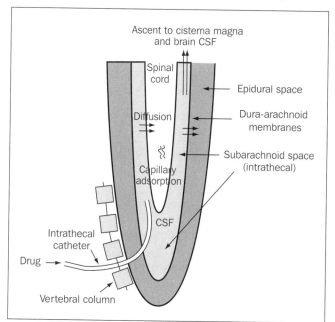

Multiple choice questions

1. **In relation to the absorption of drugs across membranes, indicate which of the following statements are correct:**
a. The percentage absorption increases linearly with increase in log P.
b. The percentage absorption decreases linearly with increase in log P.
c. There is an optimum log P for absorption.
d. Drugs with a high log P are strongly protein-bound.

2. **Good oral absorption is favoured when:**
a. The molecular weight of the drug is high.
b. Log P is less than 5.
c. There are more than 5 hydrogen bond donors.
d. There are fewer than 10 hydrogen bond acceptors.

3. **Indicate which of the following statements is correct. According to the pH-partition hypothesis:**
a. Only ionised drugs pass through membranes.
b. Weakly acidic drugs are absorbed well when the pH is below their pK_a.
c. Weakly basic drugs are absorbed well when the pH is below their pK_a.
d. Drug absorption is not usually affected by ionisation of the drug molecule.

4. **In relation to absorption from the gastrointestinal tract, which of the following statements are correct?**
a. The stomach is the main site of absorption in the gastrointestinal tract.
b. The absorbing area of the large intestine is enlarged by the presence of villi.
c. The stomach volume varies with the content of food.
d. Bile salts are secreted into the jejunum.
e. The stomach empties solids faster than liquids.
f. Natural triglycerides inhibit gastric motility.
g. The large intestine is mainly concerned with the absorption of water.

5. **In relation to buccal and sublingual absorption, which of the following statements are correct?**
a. Drugs absorbed by these routes bypass the liver.
b. Absorption through the buccal epithelium is not affected by the partition coefficient of the drug.
c. Buccal absorption of basic drugs increases with increasing pH of their solutions.
d. Buccal absorption of acidic drugs increases with increasing pH of their solutions.
e. There is an optimum log P for sublingual absorption.

6. **In relation to absorption of drugs from i.m. and s.c. injections, which of the following statements are correct?**
a. Dispersion of soluble drugs from the injection site is more rapid the lower the molecular weight of the drug.
b. Binding to muscle protein increases the rate of absorption.
c. Hydrophilic drugs bind strongly to muscle protein.
d. Muscle tissue is more acidic than normal physiological fluids.
e. Oily vehicles may be used to provide diffusion over a prolonged period.

7. **In relation to transdermal delivery of drugs, which of the following statements are correct?**
a. The main barrier of the skin is the *stratum corneum*.
b. Solute molecules may penetrate the skin through sweat ducts.
c. The major absorption pathway for lipid-soluble molecules is the transcellular route.
d. Passage of drugs through damaged skin is more rapid than through normal skin.
e. There is an optimal partition coefficient for absorption.

8. **In relation to the absorption of drugs from the eye, which of the following statements are correct?**
a. The absorbing surface is the cornea.
b. Eye drops are usually formulated to be isotonic with tears.
c. Ionised drugs are more readily absorbed than unionised drugs.
d. The rate of drainage of eye drops decreases as their viscosity decreases.
e. Most of a drug applied in the form of eye drops acts on ocular tissues.

9. **In relation to inhalation therapy, which of the following statements are correct?**
a. The pulmonary epithelium has a relatively high permeability to water-soluble molecules compared to the gastrointestinal mucosa.
b. Particles of hygroscopic materials are less readily removed from the air stream than non-hygroscopic particles.
c. Lipid-soluble molecules are less readily absorbed than lipid-insoluble molecules.
d. Only very fine particles are able to reach the alveoli.

10. **In relation to rectal absorption of drugs, which of the following statements are correct?**
a. The melting point of the suppository base is usually increased when a drug is dissolved in it.
b. Administration of drugs rectally ensures that the liver is bypassed.
c. The rate-limiting step in drug absorption from suppositories is the rate of solution of the drug in the body fluids.
d. Water-soluble drugs are more readily absorbed from a fatty excipient than from a water-soluble one.
e. Addition of surfactants to the suppository may increase the extent of absorption.

chapter 8
Physicochemical drug interactions and incompatibilities

Overview

- This chapter deals with drugs and their interactions with each other, with solvents and with excipients in formulations.
- The topic is discussed from a physicochemical rather than a pharmacological or pharmacodynamic viewpoint.
- Interactions can be beneficial, but more often they are to be avoided.

Solubility problems

- Some drugs designed to be administered by the intravenous route cannot safely be mixed with all available intravenous fluids because of poor solubility and resultant crystallisation of the drug.
- If the solubility of a drug in a particular infusion fluid is low, crystallisation may occur (sometimes very slowly) when the drug and fluid are mixed. Microcrystals may be formed which are not immediately visible.
- The mechanism of crystallisation from solution will often involve a change in pH.
- The pH of commercially available infusion fluids can vary within a range of 1–2 pH units. Therefore a drug may be compatible with one batch of fluid but not another.
- The application of the equations relating pH and pK_a to solubility should allow drug additions to be safely made or to be avoided.

KeyPoints

- Drug–drug or drug–excipient interactions can take place before administration of a drug. These may result in precipitation of the drug from solution, loss of potency or instability.
- An *incompatibility* occurs when one drug is mixed with other drugs or agents, producing a product unsuitable for administration. Reasons might be some modification of the effect of the active drug, such as an increase in toxicity or decrease in activity through some physical change such as decrease in solubility or stability.
- There are several causes of interactions and incompatibilities, which include:
  - changes in pH which may lead to precipitation of the drug
  - change of solvent characteristics on dilution, which may also cause precipitation
  - cation–anion interactions in which complexes are formed
  - the influence of salts on decreasing or increasing solubility, respectively *salting-out* and *salting-in*
  - chelation – in which a chelator binds with a metal ion to form a complex
  - ion exchange interactions where ionised drugs interact with oppositely charged resins
  - adsorption to excipients and containers causing loss of drug
  - interactions with plastics and loss of drug
  - protein binding to plasma proteins through which the free plasma concentration of drugs is reduced.

Tip

Remember that the solubility of an ionisable drug is strongly influenced by the pH of the solution because of the effect of pH on the ionisation of the drug. Undissociated drugs cannot interact with water molecules to the same extent as ionised drugs, which are readily hydrated and therefore more soluble. A change of pH can therefore sometimes lead to precipitation of ionised drugs. See Chapter 2 for equations linking pH to solubility.

Tip

Note that, whereas the pH indicates the concentration of hydrogen ions, some ions may be locked into the system and not free. The *titratable* acidity or alkalinity of a system may be more important than pH itself in determining compatibility and stability. For example, solutions of dextrose may have a pH as low as 4.0, but the titratable acidity in such an unbuffered solution is low, and thus the addition of a drug such as benzylpenicillin sodium or the soluble form of an acidic drug whose solubility will be reduced at low pH may not be contraindicated.

pH effects in vitro and in vivo

In vitro pH effects

- pH changes often follow from the addition of a drug substance or solution to an infusion fluid. An increase or decrease in pH may then produce physical or chemical changes in the system:
  - For example, as little as 500 mg of ampicillin sodium may raise the pH of 500 cm^3 of some fluids to over 8, and carbenicillin or benzylpenicillin may raise the pH of 5% dextrose or dextrose saline to 5.6 or even higher. Both drugs are, however, stable in these conditions.
- Chemical, as well as physical, instability may result from changes in pH, buffering capacity, salt formation or complexation.
- Chemical instability may give rise to the formation of inactive or toxic products.

In vivo pH effects

Gastric effects

- Fluids have a pH of 1–3 in normal subjects but the measured range of pH values in the human stomach is wide (up to 7).
- Changes in the acid–base balance have a marked influence on the absorption and thus on the activity of drugs.
- Ingestion of antacids, food and weak electrolytes will change the pH of the stomach.
- Antacids also have an effect on gastric emptying rate. Gastric emptying tends to become more rapid as the gastric pH is raised, but antacid preparations containing aluminium or calcium can retard emptying. Magnesium preparations promote gastric emptying.

Intestinal absorption

- The pH gradients between the contents of the intestinal lumen and capillary blood are smaller than in the stomach.
- Sudden changes in the acid–base balance will, nonetheless, change the concentration of drugs able to enter cells, although pH changes can change binding of the drug to protein, or drug excretion.

The importance of urinary pH

- Ingestion of some antacids over a period of 24 h will increase urinary pH and hence affect renal resorption and handling of the drug.

- Administration of sodium bicarbonate with aspirin reduces blood salicylate levels by about 50%, due to increased salicylate excretion in the urine.
- When a drug is in its unionised form it will more readily diffuse from the urine to the blood.
- Change in urinary pH will change the rate of urinary drug absorption (Figure 8.1).
- In acidic urine, acidic drugs will diffuse back into the blood from the urine.
- Acidic compounds such as nitrofurantoin are excreted faster when the urinary pH is alkaline.

Precipitation of drugs in vivo

- Pain on injection may be the result of precipitation of a drug at the site of injection brought about by either solvent dilution or by alteration in pH.
- Precipitation of drugs from formulations used intravenously can lead to thromboembolism.
- The kinetics of precipitation under realistic conditions must be taken into account; if the rate of infusion is sufficiently slow, precipitated drug may redissolve and so this problem is avoided.
- The flow rate of blood or normal saline (Q) required to maintain a drug in solution during its addition to an intravenous fluid can be predicted from $Q = R/S_m$ where R is the rate of injection of drug in mg min^{-1} and S_m is the drug's apparent maximum solubility in the system (mg cm^{-3}).

Figure 8.1 Schematic representation of the influence of urinary pH on the passive reabsorption of a weak acid and a weak base from the urine.

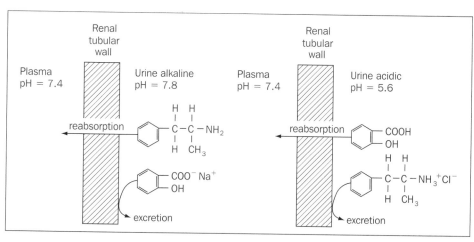

Dilution of mixed solvent systems

- Care should be taken when injectable products containing, as examples, phenytoin, digoxin and diazepam, formulated in a non-aqueous, water-miscible solvent (such as an alcohol–water mixture) or as a solubilised (e.g. micellar) preparation are diluted in an aqueous infusion fluid.
- Addition of the formulation to water may result in precipitation of the drug, depending on the final concentration of the drug and solvent.
- When a drug dissolved in a cosolvent system is diluted with water, both drug and cosolvent are diluted. The *logarithm* of the solubility of a drug in a cosolvent system generally increases linearly with the percentage of cosolvent present (Figure 8. 2). On dilution, the drug concentration falls *linearly* (not logarithmically) with a fall in the percentage of cosolvent. When the drug concentration is high the system may become supersaturated on dilution, causing precipitation.

Figure 8.2 Dilution profiles of solutions (I, II and III) containing 1, 2 and 3 mg cm⁻³ of drug respectively plotted on a semilog scale along with the solubility line. Reproduced from S.H. Yalkowsky and Valvani S., *Drug Intell. Clin. Pharm.* 11, 417 (1997).

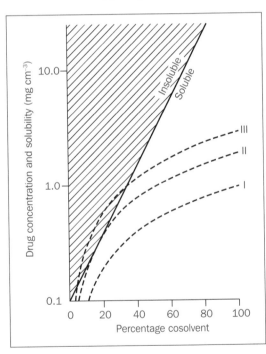

Tip

Interaction between ionised drugs will form complexes with reduced (or zero) charge. It is therefore not surprising that there will be a reduction in solubility, often leading to precipitation.

Cation–anion interactions

- The interaction between a large organic anion and an organic cation may result in the formation of a relatively insoluble precipitate.

- Complexation, precipitation or phase separation can occur in these circumstances; the product is affected by changes in ionic strength, temperature and pH.
- Examples of cation–anion interactions include those between the following drugs:
- Procainamide and phenytoin sodium, procaine and thiopental sodium, hydroxyzine hydrochloride and benzylpenicillin.
- Nitrofurantoin sodium and alkyl *p*-hydroxy benzoates (parabens), phenol or cresol, all of which tend to precipitate the nitrofurantoin.

Ion-pair formation

- Ion-pair formation may be responsible for the absorption of highly charged drugs such as the quaternary ammonium salts and sulfonic acid derivatives, the absorption of which is not explained by the pH-partition hypothesis.
- Why? The formation of an ion pair results in the 'burying' of the charges (Figure 8.3).
- Ion pairs may be considered to be neutral species formed by electrostatic attraction between oppositely charged ions in solution.
- They are often sufficiently lipophilic to dissolve in non-aqueous solvents.

KeyPoints

- Complexes which form are not always fully active, nor is their formation obvious from the ingredients. For example, neomycin sulfate *unexpectedly* forms a complex when incorporated into Aqueous Cream BP. This is because the Aqueous Cream BP comprises 30% emulsifying ointment which itself contains 10% sodium lauryl sulfate or similar anionic surfactant. The complex that forms is between neomycin sulfate and the anionic surfactant.
- Interactions are not always visible. The formation of visible precipitates depends to a large extent on the insolubility of the two combining species in the particular mixture and the size to which the precipitated particles grow.
- Interactions between drugs and ionic macromolecules are another potential source of problems:
- Heparin sodium and erythromycin lactobionate are contraindicated in admixture, as are heparin sodium and chlorpromazine hydrochloride or gentamicin sulfate.
- The activity of phenoxymethylpenicillin against *Staphylococcus aureus* is reduced in the presence of various macromolecules such as acacia, gelatin, sodium alginate and tragacanth.

Figure 8.3 Representation of an organic ion pair.

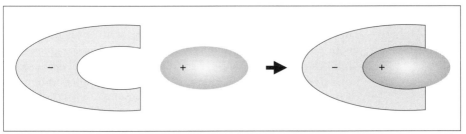

Tip

Chelation and other forms of complexation

- The term *chelation* (derived from the Greek *chele*, lobster's claw) relates to the interaction between a metal atom or ion and another species, known as the *ligand*, by which a heteroatomic ring is formed.
- Chelation changes the physical and chemical characteristics of the metal ion, and the ligand.
- It is simplest to consider the ligand as the electron pair donor and the metal as the electron pair acceptor, with the donation establishing a coordinate bond (see, for example, the copper–glycine chelate).

Structure I The 1:1 copper–glycine chelate.

- Many chelating agents act in the form of anions which coordinate to a metal ion. For chelation to occur there must be at least two donor atoms capable of binding to the same metal ion, and ring formation must be sterically possible. For example, ethylenediamine (1,2 diaminoethane, $NH_2CH_2CH_2NH_2$) has two donor nitrogens and acts as a bidentate (two-toothed) ligand.
- When a drug forms a metal chelate the solubility and absorption of both drug and metal ion may be affected, and drug chelation can lead to either increased or decreased absorption.
- Therapeutic chelators are used in syndromes where there is metal ion overload. For example, ethylenediaminetetraacetic acid (EDTA) as the monocalcium disodium salt is used in the treatment of lead poisoning; the calcium averts problems of calcium depletion. Deferiprone chelates iron.
- *Tetracycline chelation* with metal ions is a widely quoted example of complex formation leading to decreased drug absorption:
  - Polyvalent cations such as Fe^{2+} and Mg^{2+}, and anions such as the trichloracetate or phosphate interfere with absorption in both model and real systems. Ferrous sulfate has the greatest inhibitory effect on tetracycline absorption perhaps because it dissolves in water more quickly than organic iron compounds.
  - All the active tetracyclines form stable chelates with Ca^{2+}, Mg^{2+} and Al^{3+}.

- The antibacterial action of the tetracyclines depends on their metal-binding activity, as their main site of action is on the ribosomes, which are rich in magnesium.
- Tetracyclines readily form complexes with divalent metals, but they have a greater affinity for the trivalent metals with which they form 3:1 drug–metal chelates.
- Therapeutically active tetracyclines form 2:1 complexes with cupric, nickel and zinc ions while inactive analogues form only 1:1 complexes.
- The site of chelation is the C_{11}, C_{12} enolic system on the tetracycline molecule (Structure II); isochlortetracycline, which lacks this, does not chelate with Ca^{2+} ions.

Structure II Tetracycline.

- The highly coloured nature of tetracycline chelates such as the uranyl ion–tetracycline complex may be utilised in analytical procedures.

Other types of complex
- Molecular complexes of many types may be observed in systems containing two or more drug molecules:
- Generally, association follows from attractive interactions (hydrophobic, electrostatic or charge transfer interactions) between two molecules.
- The imidazole moiety is involved in many interactions. For example, caffeine and theophylline are frequently implicated in interactions with aromatic species. Caffeine increases the solubility of ergotamine and benzoic acid.

Ion exchange interactions
- Ion exchange resins are used medicinally and as systems for modified release of drugs.
- Colestyramine and colestipol are insoluble quaternary ammonium anion exchange resins which, when administered orally, bind bile acids and increase their elimination because the high-molecular-weight complex is not absorbed.
- As bile acids are converted in vivo into cholesterol, colestyramine is used as a hypocholesteraemic agent. When given to patients receiving other drugs as well, the resin

would conceivably bind anionic compounds and reduce their absorption. For example, phenylbutazone, warfarin, chlorothiazide and hydrochlorothiazide are strongly bound to the resin in vitro.

– Decreased drug absorption can be caused by use of colestyramine or colestipol and has been reported, for example, with thyroxine, aspirin, phenprocoumon, warfarin, chlorothiazide, cardiac glycosides and ferrous sulfate.

Adsorption of drugs

▪ Adsorbents can be used to remove noxious substances from the lumen of the gut.

▪ Unfortunately, they are generally non-specific so will also adsorb nutrients, drugs and enzymes when given orally.

▪ Several consequences of adsorption are possible:

– If the drug remains adsorbed until the preparation reaches the absorption site, the concentration of the drug presented to the absorbing surfaces will be much reduced. The driving force for absorption would then be reduced, resulting in a slower rate of absorption.

– Alternatively, the release of drug from the adsorbent might be complete before reaching the absorption site, possibly hastened by the presence of electrolytes in the gastrointestinal tract, in which case absorption rates would be virtually identical to those in the absence of adsorbent.

– Loss of activity of preservatives can arise from adsorption on to solid drug substances. Benzoic acid, for example, can be adsorbed to the extent of 94% by sulfadimidine.

Protein and peptide adsorption

▪ Adsorption of peptides to glass or plastic may occur because of the amphipathic nature of many peptides. This becomes pharmaceutically important when they are originally present in low concentrations in solution.

▪ The adsorption of peptides onto glass is ascribed to bonding between their amino groups and the silanol groups of the glass.

Drug interactions with plastics

▪ The plastic tubes and connections used in intravenous containers and giving sets can adsorb or absorb a number of drugs leading to significant losses in some cases.

▪ Those drugs which show a significant loss when exposed to plastic, in particular poly(vinyl chloride), include insulin, nitroglycerin, diazepam, clomethiazole, vitamin A acetate, isosorbide dinitrate and a miscellaneous group of drugs such as phenothiazines, hydralazine hydrochloride and thiopental sodium.

- Preservatives such as the methyl and propyl parabens present in formulations can be sorbed into rubber and plastic membranes and closures, thus leading to decreased levels of preservative and, in the extreme, loss of preservative activity.

Tips

Note the difference between *adsorption* and *absorption*:
- Adsorption is the attachment of a molecule to a surface.
- Absorption involves its penetration into the substance with which it interacts.

Protein binding of drugs

Binding of drugs to proteins is important because the bound drug assumes the diffusional and other transport characteristics of the protein molecule:

- For example, drugs bound to albumin (or other proteins) are attached to a unit too large to be transported across membranes. They are thus prevented from reacting with receptors or from entering the sites of drug metabolism or drug elimination, until they dissociate from the protein.

A second important consequence of protein binding is that the free drug concentration is reduced. This is important because it is only free drug that is able to cross the capillary endothelium:

- In cases where drug is highly protein-bound (around 90%), small changes in binding lead to drastic changes in the levels of free drug in the body.
- Both ampicillin (50 mg kg^{-1} every 2 h) and oxacillin (50 mg kg^{-1} h^{-1}) produce similar peak levels in the serum given as repeated intravenous boluses. Levels of free drug are markedly different, however, as oxacillin is 75% protein-bound and ampicillin is 17.5% bound.
- The level of free drug in serum is important in determining the amount of drug that reaches tissue spaces because it determines the gradient of drug concentration between the serum and the tissues. This relationship is given by:

$$C_t = \frac{C_s f_s}{f_t}$$

where C_t is the total concentration of drug in tissue fluid, C_s is the serum drug concentration, and f_s and f_t are the free fractions of drug in serum and tissue fluid, respectively.

Binding to plasma proteins

- Most drugs bind to a limited number of sites on the albumin molecule.
- Plasma proteins other than albumin may also be involved in binding. Blood plasma normally contains on average about 6.72 g of protein per 100 cm³; the protein comprises 4.0 g of albumin, 2.3 g of globulins and 0.24 g of fibrinogen.
- Dicoumarol binds to β- and γ-globulins, and certain steroid hormones are specifically and preferentially bound to particular globulin fractions.
- Binding to plasma albumin is generally easily reversible, so that drug molecules bound to albumin will be released as the level of free drug in the blood declines.
- Binding to albumin is a process comparable to partitioning of drug molecules from a water phase to a non-polar phase. The hydrophobic sites, however, are not necessarily 'preformed'.

Lipophilicity and protein binding

- The extent of protein binding of many drugs is a linear function of their partition coefficient P (or log P).
- A linear equation of the form:
  log (percentage bound/percentage free) = 0.5 log P – 0.665
  may be applied to serum binding of penicillins. Although there may be an electrostatic component to the interaction, the binding increases with the degree of lipophilicity, suggesting, as is often the case, that more than one binding interaction is in force.

Muscle protein may bind drugs such as digoxin and so act as a depot. Concentrations of 1.2 ± 0.8, 11.3 ± 4.9 and 77.7 ± 43.3 ng cm⁻³ have been reported for digoxin in plasma, skeletal and cardiac muscle, respectively.

- Dicloxacillin, which is 95% bound to protein, is absorbed more slowly from muscle than ampicillin, which is only bound to the extent of 20%.

Protein binding can affect antibiotic action. For example:

- Penicillins and cephalosporins bind reversibly to albumin. Only the free antibiotic has antibacterial activity.
- Oxacillin in serum at a concentration of 100 µg cm⁻³ exhibits an antibacterial effect similar to that of 10 µg cm⁻³ of the drug in water. A high degree of serum protein binding may nullify the apparent advantage of higher serum levels of some agents.

The degree of binding of drug D to protein P may be estimated as follows:

- Assuming that protein binding can be considered to be an adsorption process obeying the law of mass action:
$$D + P \rightleftharpoons (DP)$$
(DP = protein–drug complex)

- Then, at equilibrium:

$$D_f + (P_t - D_b) = D_b$$

where D_f is the molar concentration of unbound drug, P_t is the total molar concentration of protein and D_b is the molar concentration of bound drug (= molar concentration of complex).

- The ratio r of the number of moles bound to the total protein in the system can be shown to be:

$$r = \frac{nKD_f}{1 + KD_f}$$

or:

$$\frac{1}{r} = \frac{1}{n} + \frac{1}{nKD_f}$$

where n is the number of binding sites per molecule and K is the ratio of the rate constants for association and dissociation.

- The fraction of drug-bound β generally varies with the concentration of both drug and protein and is given by:

$$\beta = \frac{1}{1 + D_f/(nP_t) + 1/(nKP_t)}$$

Multiple choice questions

1. **Indicate which of the following statements are true. In general, the administration of antacids would be expected to:**
a. increase the gastric pH
b. affect gastric emptying time
c. decrease urinary pH
d. decrease the ionisation of acidic drugs in the stomach
e. reduce the gastric absorption of acidic drugs

2. **Indicate which of the following statements are true. Acidification of the urine:**
a. may result from the administration of antacids
b. would be expected to increase the rate of urinary excretion of acidic drugs such as nitrofurantoin
c. would be expected to increase the rate of urinary excretion of basic drugs such as imipramine
d. may cause precipitation of drugs in the urine
e. would be expected to increase the reabsorption of a basic drug

3. **Indicate which of the following statements are true. Ion pair formation:**
a. occurs between ions of similar charge
b. is a consequence of electrostatic interaction between ions
c. decreases the absorption of quaternary ammonium salts
d. results in the formation of a neutral species

4. **Indicate which of the following statements relating to the formation of chelates are true:**
a. Chelation is the interaction between a large organic anion and an organic cation.
b. Chelation is the interaction between a metal ion and a ligand.
c. The ligand is the electron pair acceptor.
d. Chelation results in the formation of a heteroatomic ring.
e. Tetracyclines form 3:1 drug–metal chelates with magnesium ions.

5. **Indicate which of the following interactions might be expected to lead to an increased drug absorption:**
a. ion pair formation
b. drug binding to plasma proteins
c. adsorption of drug on to antacids
d. interaction of drug with the plastic tubing of giving sets
e. chelation of tetracyclines with metal ions

6. **Indicate which of the following statements relating to the protein binding of drugs are true:**
a. Protein binding to plasma albumin is a reversible process.
b. Protein binding decreases the free drug concentration.
c. Drugs with a low lipophilicity have a high degree of protein binding.
d. Protein binding of antibiotics usually increases antibiotic action.

7. **If the rate of injection of a drug is 7 mg min^{-1}, and the maximum solubility of the drug is approximately 0.5 mg cm^{-3}, what blood or saline flow rate is required to prevent observable precipitation?**
a. 14 cm^3 min^{-1}
b. 2 cm^3 min^{-1}
c. 3.5 cm^3 min^{-1}
d. 0.29 cm^3 min^{-1}

8. **A drug is bound to serum albumin to the extent of 95% bound. What is the percentage effect on the free levels of the drug of a reduction in binding to 92%?**
a. 3% increase
b. 8% increase
c. 160% increase
d. 3% decrease

chapter 9

Peptides, proteins and other biopharmaceuticals

Overview

This chapter discusses:

- some of the basic properties of peptides and proteins and how their physical properties are dictated not only by the properties of their individual amino acids but also by the spatial arrangement of the amino acids in their polypeptide chains
- the physical and chemical stability of protein pharmaceuticals and formulation procedures for their stabilisation
- the formulation and properties of some therapeutic proteins and peptides and of DNA

Structure and solution properties of peptides and proteins

Definitions

- *Peptide*: a short chain of amino acid residues with a defined sequence (e.g. leuprolide).
- *Protein*: polypeptides which occur naturally and have a defined sequence of amino acids and a three-dimensional structure (e.g. insulin).

Structure of peptides and proteins

Proteins have in increasing order of complexity (Figure 9.1):

- *Primary* structure – the order in which the individual amino acids are arranged.
- *Secondary* structures – including coiled α-helix and pleated sheets.
- *Tertiary* structure – the three-dimensional arrangement of helices and coils.
- *Quaternary* forms – the association of ternary forms (e.g. the hexameric form of insulin).

KeyPoints

- Most peptides and proteins are not absorbed to any significant extent by the oral route and most available formulations of protein pharmaceuticals are therefore parenteral products for injection, or inhalations.
- DNA, RNA and various oligonucleotides are increasingly used in gene therapy. These share some of the problems of proteins as therapeutic agents.

KeyPoint

Loss of the unique tertiary or quaternary structure, through denaturation, can occur from a variety of insults that would not affect smaller organic molecules. Formulations must preserve the protein structure.

Hydrophobicity of peptides and proteins

- Amino acids have a range of physical properties, each having a greater or lesser degree of hydrophilic or hydrophobic nature.
- If amino acids are spatially arranged in a molecule so that distinct hydrophobic and hydrophilic regions appear, then the polypeptide or protein will have an amphiphilic nature.

Figure 9.1 Diagrammatic representation of protein structure.

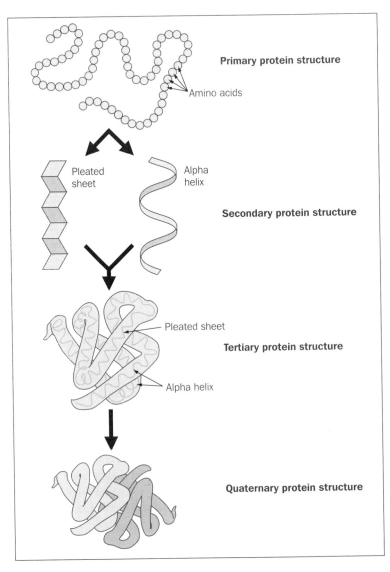

Definition

■ *Polypeptide*: a longer amino acid chain, usually of defined sequence and length (e.g. vasopressin).

Solubility of peptides and proteins

■ The aqueous solubilities of proteins vary enormously, from the very soluble to the virtually insoluble.
■ The solubility of globular proteins increases as the pH of the solution moves away from the *isoelectric point* (IP), which is the pH at which the molecule has a net zero charge (Figure 9.2).
■ At its IP a protein has a tendency to self-associate.
■ As the net charge increases, the affinity of the protein for the aqueous environment increases and the protein molecules also exert a greater electrostatic repulsion.
■ Proteins are surrounded by a hydration layer, equivalent to about 0.3 g H_2O per gram of protein (about 2 water molecules per amino acid residue).
■ Aqueous solutions of proteins sometimes exhibit phase transitions (Figure 9.3). The phase behaviour of protein solutions is affected by pH and ionic strength.
■ Addition of electrolytes such as NaCl, KCl and $(NH_4)_2SO_4$ decreases solubility.
■ At high ionic strengths proteins precipitate – a *salting-out* effect.
■ Organic solvents tend to decrease the solubility of proteins by lowering solvent dielectric constant.

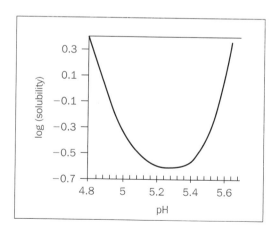

Figure 9.2 A plot of the logarithm of aqueous solubility of β-lactoglobulin versus pH.

Figure 9.3 Phase diagram for aqueous solutions of γ-crystallin.

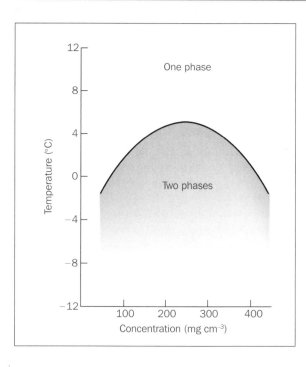

The stability of proteins and peptides

Physical instability

Denaturation

- is the disruption of the tertiary and secondary structure of the protein molecule.
- can be *reversible* or *irreversible*:
- It is reversible if the native structure is regained, for example on decreasing the temperature when temperature has caused the initial changes.
- It is irreversible when the unfolding process is such that the native structure cannot be regained.

KeyPoints

Protein pharmaceuticals can suffer both physical and chemical instability (Figure 9.4):
- *Physical instability* results from changes in the higher-order structure (secondary and above).
- *Chemical instability* is modification of the protein via bond formation or cleavage.

Aggregation

- Some proteins self-associate in aqueous solution to form oligomers.
- Insulin, for example, exists in several states:
- The zinc hexamer of insulin is a complex of insulin and zinc which slowly dissolves into dimers and eventually monomers following subcutaneous administration, conferring on it long-acting properties.

Figure 9.4 Physical and chemical pathways of protein degradation.

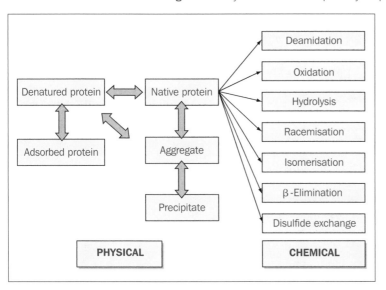

Surface adsorption and precipitation

- Adsorption of proteins such as insulin on surfaces such as glass or plastic in giving sets:
- – can reduce the amount of agent reaching the patient
- – can lead to further denaturation, which can then cause precipitation and the physical blocking of delivery ports in protein pumps.
- Denaturation is facilitated by the presence of a large head space allowing a greater interaction of proteins with the air–water interface.

Formulation and protein stabilisation
Stability testing

- Stability testing of protein-containing formulations often involves subjecting the solutions to shaking for several hours and the subsequent assessment of the protein configuration.

Improving the physical stability of proteins through formulation
Prevention of adsorption

- Additives can coat the surface of glass or bind to the proteins.
- Serum albumin can be included in the formulation to compete with the therapeutic protein for the binding sites on glass and reduce adsorption.
- A similar effect can be achieved by the addition of surfactants such as poloxamers and polysorbates to the protein solution.

Minimisation of exposure to air

- ▣ Significant denaturation of proteins can occur when the protein solutions are exposed at the air–solution interface.
- ▣ Agitation of protein solutions in the presence of air or application of other shear forces (e.g. in filters or pumps) may lead to denaturation.
- ▣ The inclusion of surfactants can reduce denaturation arising from these processes.

Addition of cosolvents

- ▣ Some excipients and buffer components added to the protein solution are able to minimise denaturation through their effects on solvation.
- ▣ These include polyethylene glycols and glycerol, referred to as *cosolvents.*
- ▣ These act either by causing the *preferential hydration* of the protein or alternatively by *preferential binding* to the protein surface (Figure 9.5):
- – Preferential hydration results from an exclusion of the cosolvent from the protein surface due to steric effects (as in the case of polyethylene glycols) or surface tension effects (as with sugars, salts and amino acids). As a result more water molecules pack around the protein in order to exclude the additive and the protein becomes fully hydrated and stabilised in a compact form.

Figure 9.5 Diagram showing preferential binding and preferential hydration by solvent additives. Reproduced from Timasheff SN, Arakawa T (1989) In: Creighton T E (ed.) *Protein Structure: A Practical Approach*. Oxford: IRL Press, pp 331–345.

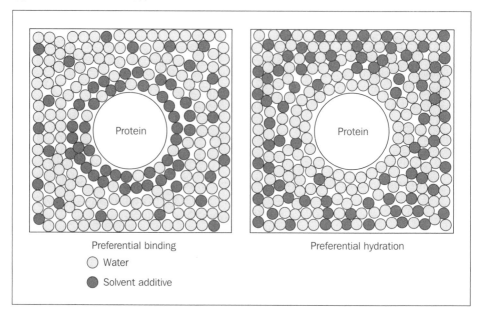

Preferential binding   Preferential hydration

○ Water
● Solvent additive

- Alternatively, the cosolvent may stabilise the protein molecule by preferentially binding to it either non-specifically or to specific sites on its surface.

Optimimisation of pH

- To avoid stability problems arising from charge neutralisation and to ensure adequate solubility, a pH must be selected which is at least 0.5 pH units above or below the IP.
- Since a pH range of 5–7 is usually required to minimise chemical breakdown, this frequently coincides with the IP.

Characterisation of degradation

- If the formulation does not prevent denaturation and aggregation of the protein, then the pharmacology, immunogenicity and toxicology of the denatured or aggregated protein must be studied to determine its safety and efficacy.
- If the aggregates are soluble there may be a significant effect on the pharmacokinetics and immunogenicity of the protein.
- Insoluble aggregates are generally unacceptable.

Chemical instability

Deamidation

In deamidation the side-chain linkage in a glutamine (Gln) or asparagine (Asn) residue is hydrolysed to form a free carboxylic acid.

Prevention of deamidation

- If the deamidation occurs by a general acid–base mechanism then the optimum pH for a peptide formulation will usually be about 6, where both rates are at their minimum.
- If the deamidation occurs through the cyclic imide intermediate it is preferable to formulate at a low pH since this type of deamidation is base-catalysed.

Oxidation

- Oxidation is one of the major causes of protein degradation.
- The side chains of histidine (His), methionine (Met), cysteine (Cys), tryptophan (Trp) and tyrosine (Tyr) residues in proteins are potential oxidation sites.
- Methionine is very susceptible to oxidation and reacts with a variety of oxidants to give methionine sulfoxide ($RS(OO)CH_3$) or, in highly oxidative conditions, methionine sulfone ($RS(O)CH_3$).
- The thiol group of cysteine readily reacts with oxygen to yield, successively, sulfenic acid (RSOH), a disulfide (RSSH), a sulfinic acid (RSO_2H) and, finally, a sulfonic (cystic) acid (RSO_3H) depending on reaction conditions.

- An important factor determining the extent of oxidation is the spatial positioning of the thiol groups in the proteins.
- Histidine is susceptible to oxidation in the presence of metals, primarily by reaction with singlet oxygen, and this constitutes a major cause of enzyme degradation.
- Both histidine and tryptophan are highly susceptible to photooxidation.

Prevention of oxidation

- In most cases oxidation results in a complete or partial loss of activity.
- Minimising protein oxidation is essential for maintaining the biological activity of most proteins and avoiding the immunogenic response caused by degraded proteins.
- A variety of measures may be employed in order to prevent protein oxidation:
- temperature reduction, either by refrigeration or by freezing
- control of pH if the rate of oxidation is pH-dependent.

Racemisation

All amino acid residues except glycine (Gly) are chiral at the carbon atom bearing the side chain and are subject to base-catalysed racemisation.

Proteolysis

Proteolysis involves the cleavage of peptide (–CO–NH–) bonds:
- Asp is the residue most susceptible in proteolysis.
- The cleavage of the peptide bonds in dilute acid proceeds at a rate at least 100 times that of other peptide bonds.

Beta-elimination

High-temperature treatment of proteins leads to destruction of disulfide bonds as a result of β-elimination from the cystine residue:
- The inactivation of proteins at high temperatures is often due to β-elimination of disulfides from the cystine residue.
- Other amino acids, including Cys, Ser, Thr, Phe and Lys, can also be degraded via β-elimination.
- The inactivation is particularly rapid under alkaline conditions and is also influenced by metal ions.

Disulfide formation

The interchange of disulfide bonds can result in incorrect pairings with consequent changes of three-dimensional structure and loss of catalytic activity.

Accelerated stability testing of protein formulations

- The mechanisms of degradation at higher temperatures may not be the same as at lower temperatures and the application of the Arrhenius equation in the prediction of protein stability will be more uncertain than with small-molecule drugs.
- Deviation from the Arrhenius equation occurs, however, if the protein exists in multiple conformational forms that retain activity during unfolding.

Protein formulation and delivery

Protein and peptide transport

- It can be shown that, in an in vitro intestinal cell monolayer system, a good correlation is found between the permeability coefficient, P, and the log of the partition coefficient of the peptides between heptane and ethylene glycol (rather than octanol and water).
- Molecular volume (or size) will increasingly be a factor influencing transport as the molecular weight of the peptide increases.
- The rate of translational movement depends on the size of the molecule, its shape and interactions with solvent molecules.

Lyophilised proteins

- Proteins such as insulin, tetanus toxoid, somatotropin and human albumin aggregate in the presence of moisture, which can lead to reduced activity, stability and diffusion.
- Because of their potential instability in solution, therapeutic proteins are often formulated as lyophilised powders.
- Even in this state several suffer from moisture-induced aggregation.

Controlled delivery of proteins and peptides

A wide range of biodegradable polymers are used for the controlled delivery of proteins and peptides, including:

- natural substances, starch, alginates, collagen
- a variety of proteins such as cross-linked albumin

■ a range of synthetic hydrogels, polyanhydrides, polyesters or orthoesters, poly(amino acids) and poly(caprolactones). Poly(lactide-glycolide) is one of the commonest polymers used in microsphere form to deliver proteins and peptides.

Definition

■ *Polyamino acids*: random sequences of varying lengths, generally resulting from non-specific polymerisation of one or more amino acids (e.g. glatiramer).

Routes of delivery

Invasive and non-invasive routes of delivery for peptides and proteins involve direct injection of solutions, depot systems and a variety of nasal, inhalation, topical and other formulations.

Therapeutic proteins and peptides

Some examples of therapeutic proteins and peptides, including their size and use or action, are given in Table 9.1.

Table 9.1 Some therapeutic proteins and peptides, their molecular weights and actions

Protein/peptide	Size (kDa)	Use/action
Oxytocin	1.0	Uterine contraction
Vasopressin	1.1	Diuresis
Leuprolide acetate	1.3	Prostatic carcinoma therapy
LHRH analogues	~1.5	Prostatic carcinoma therapy
Somatostatin	3.1	Growth inhibition
Calcitonin	3.4	Ca^{2+} regulation
Glucagon	3.5	Diabetes therapy
Parathyroid hormone (1–34)	4.3	Ca^{2+} regulation
Insulin	6	Diabetes therapy
Parathyroid hormone (1–84)	9.4	Ca^{2+} regulation
Interferon-gamma	16 (dimer)	Antiviral agent
TNF-α	17.5 (trimer)	Antitumour agent
Interferon α-2	19	Leukaemia, hepatitis therapy
Interferon β-1	20	Lung cancer therapy
Growth hormone	22	Growth acceleration
DNase	~32	Cystic fibrosis therapy
α_1-Antitrypsin	45	Cystic fibrosis therapy
Albumin	68	Plasma volume expander
Bovine IgG	150	Immunisation
Catalase	230	Treatment of wounds and ulcers
Cationic ferritin	400+	Anaemias

LHRH, luteinising hormone-releasing hormone; TNF-α, tumour necrosis factor-α; IgG, immunoglobin G. Reproduced from Niven R W. *Pharm Technol* 1993; July: 72.

Insulin

There are three main types of insulin preparation:

1. Those with a *short* duration of action which have a relatively rapid onset (soluble insulin, insulin lispro and insulin aspart).
2. Those with an *intermediate* action (isophane insulin and insulin zinc suspension).
3. Those with a slower/slow action, in onset and lasting for long periods (crystalline insulin zinc suspension).

Precipitation of insulin and other proteins

- Precipitation of insulin in pumps due to the formation of amorphous particles, crystals or fibrils of insulin can lead to changes in release pattern.
- 'Amorphous' or 'crystalline' precipitates can be caused by the leaching of divalent metal contaminants or lowering of pH (due to CO_2 diffusion or leaching of acidic substances).
- Interactions leading to fibril formation result from change in monomer conformation and hydrophilic attraction of the parallel β-sheet forms.
- Fibril formation is also encouraged by contact of the insulin solution with hydrophobic surfaces.
- Chemical modifications to an endogenous protein, however minor, can lead to significant differences in properties and activity.
- Recombinant human protein analogues may be subtly different.
- There is as yet no simple way to predict the consequence of subtle changes in structure.

Calcitonin

- Calcitonin, a peptide hormone of 32 amino acids having a regulatory function in calcium and phosphorus metabolism, is used in various bone disorders such as osteoporosis.
- Salmon, human, pig and eel calcitonin are used therapeutically.
- Species differences may be significant – salmon calcitonin is 10 times more potent than human calcitonin.
- Human calcitonin has a tendency to associate rapidly in solution and, like insulin, form fibrils, resulting in a viscous solution. The fibrils are 8 nm in diameter and often associate with one another.

DNA and oligonucleotides

DNA

- DNA of varying molecular weights (base pairs) is used in gene therapy.
- As a large hydrophilic, polyanionic and sensitive macromolecule, successful delivery to target cells and the nucleus within these cells is an issue.
- Shearing of high-molecular-weight DNA while stirring in solution can lead to breakage of the molecule.
- One approach to delivery is to complex the DNA with polymers or particles of opposite charge to produce, by condensation, more compact species.
- DNA can be condensed to nanoparticles with cationic polymers (e.g. polylysine and chitosan), cationic liposomes and dendrimers. These retain an overall positive charge and are able to transfect cells more readily than native or naked DNA.

Oligonucleotides

- Antisense oligonucleotides (used for the sequence-specific inhibition of gene expression) are polyanionic molecules with between 10 and 25 nucleotides, which resemble single-stranded DNA or RNA.
- They have a molecular weight from 3000 to 8000 and are hydrophilic, having a log P of approximately -3.5.
- Like DNA, they clearly do not have the appropriate properties for transfer across biological membranes.
- They are also sensitive to nucleases and non-specific adsorption to biological surfaces.

Multiple choice questions

1. Indicate which of the following statements are true. The solubility of globular proteins:
   a. is minimum at the IP
   b. decreases when the dielectric constant of the solvent is decreased
   c. is increased by addition of electrolytes
   d. is independent of pH
   e. is higher in organic solvents than in water

2. Indicate which of the following statements are true. Denaturation of proteins:
   a. is the disruption of the tertiary and secondary structure
   b. occurs more readily in the presence of cosolvents such as glycerol
   c. cannot be reversed
   d. can be caused by agitation of protein solutions in the presence of air
   e. is increased by the presence of surfactants in the solution

3. **Indicate which of the following statements are true. The surface adsorption of proteins:**
a. can be reduced by including serum albumin in the formulation
b. is increased when surfactants are present in the formulation
c. can be prevented by using plastic containers
d. can lead to protein denaturation
e. can be minimised by reducing the headspace in the container

4. **Indicate which of the following statements relating to the chemical instability of proteins are true:**
a. Deamidation is the hydrolysis of the side-chain linkage in a glutamine (Gln) or asparagine (Asn) residue to form a free carboxylic acid.
b. Methionine is very susceptible to oxidation.
c. Oxidation can be caused by freezing the protein solutions.
d. β-elimination leads to destruction of disulfide bonds.
e. β-elimination is particularly rapid in acidic conditions.

5. **Indicate which of the following statements relating to the properties of insulin are true. Insulin can:**
a. associate in aqueous solution to form oligomers
b. form amorphous or crystalline precipitates at high pH
c. form fibrils due to the hydrophobic attraction of the parallel β-sheet forms
d. form fibrils on contact with hydrophobic surfaces
e. be formulated as an insulin zinc suspension to give rapid onset of action

chapter 10
In vitro assessment of dosage forms

Overview

This chapter recalls:
- the basics of some in vitro tests which can be applied to pharmaceutical products
- how the effect of some of the key parameters of pharmaceutical systems, such as particle size, viscosity, adhesion or formulation in general on drug release or performance can be measured
- the importance of in vitro testing in formulation development and in batch-to-batch control but also in assessing defects in products
- that in vitro tests might be preferred to in vivo measures when there is a good correlation between in vitro and in vivo behaviour.

Dissolution testing of solid dosage forms

The rate of solution of a solid drug substance from a granule or a tablet is dependent to a large extent on its solubility in the solvent phase and its concentration in that phase.

The physicochemical factors which point to the need for dissolution testing include:
- *Low aqueous drug solubility.*
- *Poor product dissolution* – evidence from the literature that the dissolution of one or more marketed products is poor.
- *Drug particle size* – evidence that particle size may affect bioavailability.
- *The physical form of drug* – when polymorphs, solvates and complexes have poor dissolution characteristics.
- *Presence of specific excipients* which may alter dissolution or absorption.
- *Tablet or capsule coating* which may interfere with the disintegration or dissolution of the formulation.

KeyPoints

- In vitro tests provide the opportunity to make precise and reproducible release measurements to distinguish between different formulations of the same drug or the same formulation after ageing or processing changes or during production, i.e. batch-to-batch variation.
- They do not replace the need for clinical work, but an in vitro test can pinpoint formulation factors during development which are of importance in determining drug release.
- Physiological verisimilitude is not essential for validity in quality control, where reproducibility of a product is in itself a goal.

KeyPoints

- With drugs of very low solubility it is sometimes necessary to consider the use of in vitro tests which allow sink conditions to be maintained. This generally involves the use of a lipid phase into which the drug can partition: alternatively it may involve dialysis or physical replacement of the solvent phase.
- Mixed-solvent systems such as ethanol–water or surfactant systems may have to be used to enhance the solubility of sparingly soluble drugs, but some prefer the use of flow-through systems in these cases.

In vitro methods may be divided into two types:

1. *Natural convection* in which, for example, a pellet of material is suspended from a balance arm in the dissolution medium. Because there is no agitation, the conditions are not representative of in vivo conditions.

2. *Forced convection* in which a degree of agitation is introduced, so making this method more representative of in vivo conditions. Most practical methods fall into this category. There are two types of forced convection methods: those that employ non-sink conditions and those that achieve sink conditions in the dissolution medium. Figure 10.1 shows examples of simple types of forced convection methods employing non-sink conditions.

Figure 10.1 Simple types of forced convection methods of dissolution testing.

Tips

- If released drug is not removed from the dissolution medium during dissolution testing, i.e. if testing is performed under non-sink conditions, the drug concentration in this medium may, in some cases, approach saturation level and if so the rate of release of the drug will be significantly reduced (see Noyes–Whitney equation, Chapter 1).
- Sink conditions normally occur when the volume of the dissolution medium is at least 5–10 times the saturation volume.

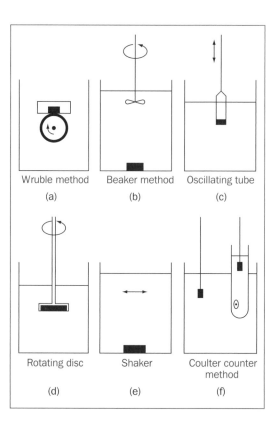

Wruble method (a)	Beaker method (b)	Oscillating tube (c)
Rotating disc (d)	Shaker (e)	Coulter counter method (f)

Experimental methods of testing tablet dissolution include:

Pharmacopoeial and compendial dissolution tests

■ The *British Pharmacopoeia* method involves a rotating wire mesh basket in which tablets or capsules are placed (Figure 10.2). The mesh is small enough to retain broken pieces of tablet but large enough to allow entry of solvent without wetting problems. The basket may be rotated at any suitable speed but most *United States Pharmocopeia* monographs specify 50, 100 or 1500 rpm.

■ In all methods the appropriate pH for the dissolution medium must be chosen and there should be a reasonable degree of agitation.

KeyPoints

■ There is no absolute method of dissolution testing.
■ Whatever form of test is adopted, results are only really useful on a comparative basis – batch versus batch, brand versus brand, or formulation versus formulation.
■ Release tests for non-oral products are less well developed than for oral products.

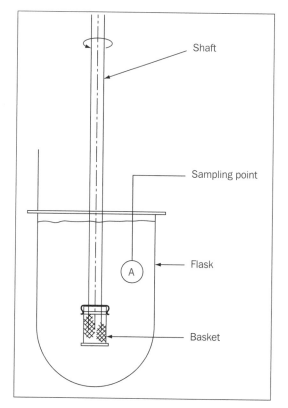

Figure 10.2 The rotating basket method.

Flow-through systems

▪ A variant of the dissolution methods discussed uses convection achieved by solvent flow through a chamber. Dissolution data obtained from such a system with continuous monitoring of drug concentration must be interpreted with care as the concentration–time profile will be dependent on the volume of solvent, its flow rate and the distance of the detection device from the flow cell, or rather the void volume of solvent.

In vitro evaluation of non-oral systems

Suppositories

▪ Suppositories are difficult to study in vitro, because it is not easy to simulate the conditions in the rectum.
▪ One system employs a suppository placed in a pH 7.8 buffer in a dialysis bag which is then placed in a second dialysis bag filled with octanol and the whole is suspended in a flow system at 37°C. The amount of drug released into the outer liquid is monitored.

In vitro release from topical products and transdermal systems

▪ In vitro testing of the lot-to-lot uniformity of semisolid dosage forms of creams, ointments and lotions is important in quality control.
▪ Ointments and transdermal systems encounter little water in use but useful data can be obtained by measuring release into aqueous media, which can sometimes be predictive of in vivo performance.
▪ Alternatively, a liquid biophase can be simulated using isopropyl myristate (Figure 10.3).
▪ A rotating bottle apparatus has been used to measure the release of nitroglycerin from Deponit transdermal patches.
▪ The *British Pharmacopoeia* specifies a distribution (release) test for transdermal patches based on the paddle apparatus for tablet and capsules.

Rheological characteristics of products

The terms 'soft and unctuous' and 'hard and stiff' are used to describe dermatologicals but are difficult to quantify.

Viscosity monitoring can be used as a quality control procedure, but some very practical rheological tests may be carried out. For example:

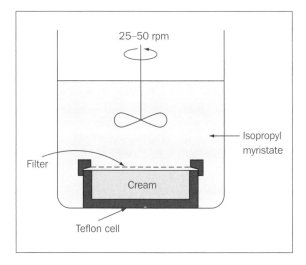

Figure 10.3 An apparatus for examination of the release of drug from a cream formulation into isopropyl myristate (*IM*). The filter is saturated with *IM*.

■ The injectability of non-aqueous injections, which are often viscous and thus difficult to inject, can be assessed by a test for syringeability. Sesame oil has a viscosity of 56 cP, but added drugs and adjuvants may increase the viscosity.

Adhesivity of dosage forms

Formulation aspects
■ Adhesive preparations have been formulated, for example, for the topical treatment of stomatitis.
■ The adhesive nature of transdermal patches is important.
■ The adhesion of film coats to tablet surfaces is a key quality issue.
■ Rubbery polymers which have partly liquid and partly elastic characteristics are employed as adhesives in surgical dressings and adhesive tapes.
■ Peeling tests for film coats are routinely used in pharmaceutical development and similar tests for the adhesion of transdermal particles to skin have been used.

Several current methods of testing oral dosage forms for adhesivity include:
■ Measurement of the force of detachment of a solid dosage form by raising the dosage form through an isolated oesophagus.
■ Assessment of the adhesion of a moistened capsule or tablet to a surface using a strain gauge. The effects of polymer concentration and composition and of additives on the adhesivity of film coating materials can be studied using this apparatus, and the force required to separate tablet from substrate measured.

The variables likely to affect the process of adhesion of coated tablets to mucosal surfaces (sometimes an unwanted effect) include:

- film coat thickness
- the nature of the film coat, for example, its hydrophobicity
- the nature of the contacting surface
- rate of coat hydration or dissolution during the adhesion process
- the rheology of the solution of film coating material formed during adhesion, its surface tension and its elongational characteristics.

KeyPoints

- Analysis of particle size distribution of aerosol formulations during formulation development, clinical trial or after storage is of obvious clinical relevance.
- Aerosols are not easy to size, primarily because they are dynamic and inherently unstable systems.

Tips

Remember that an aerosol is a type of colloidal dispersion in which the liquid or solid particles are dispersed in a continuous phase (air). According to Stokes' law:

$$v = \frac{2\,ga^2\,(\rho_1 - \rho_2)}{9\eta}$$

Consequently the rate of sedimentation (v) of the particles will *increase* with increase of the particle radius (a), and the difference between the density of the particles (ρ_1) and the continuous phase (ρ_2), and *decrease* with increase of the viscosity (η) of the continuous phase.

Particle size distribution in aerosols

Methods of sampling may be divided into:

- Techniques which utilise an aerosol cloud. Sedimentation techniques based on Stokes' law are applied and the usual detection system is photometric.
- Dynamic methods in which particles are carried in a stream of gas. Instruments utilise both sedimentation and inertial forces and depend on the properties of particles related to their mass.

The *Royco sizer* is a commercially available instrument which measures individual particles in a cloud (it is used to monitor the air of 'clean rooms').

- This instrument can be used to size particles in aerosol clouds provided that the particle size distribution does not change during the time of the analysis either by preferential settling of larger particles or by coagulation.

The *cascade impactor* is probably the most widely used instrument in categorising airborne particles. In this instrument:

- Large particles leave the airstream and impinge on baffles or on glass microscope slides (Figure 10.4).
- The airstream is then accelerated at a nozzle, providing a second range of smaller-sized particles on the next baffle and so on.

- Progressively finer particles are collected at the successive stages of impingement owing to jet velocity and decreasing jet dimension.

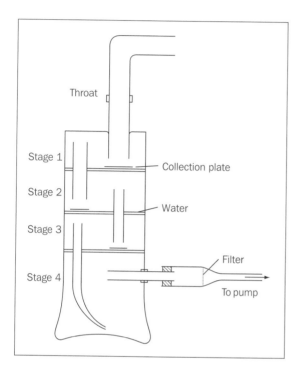

Figure 10.4 A multistage liquid impinger. Reproduced from Hallworth G W (1987). *Br. J. Clin. Pharm.* 1997; 4: 157, with permission.

'Artificial throat' devices are useful for comparative studies of the behaviour of medicinal aerosols. In these devices:
- The particles are segregated according to size.
- Analysis of the collecting layers at the several levels of the device allows the monitoring of changes in released particle size.
- Where an artificial mouth is used, washing is carried out to reveal the extent of fall-out of large particles.
- The smallest particles of all reach the collecting solvent.

The *British Pharmacopoeia* and other compendia have adopted detailed specifications for two impinger devices. These operate by dividing the dose emitted from an inhaler into the *respirable* and *non-respirable* fractions.

Apparatus A (glass) (Figure 10.5)
- Apparatus A employs the principle of liquid impingement and has a solvent in both chambers to collect the aerosol.
- Air is drawn through the system at 60 l min⁻¹ and the inhaler is fired several times into the device.

Figure 10.5 *British Pharmacopoeia* impinger apparatus A. Reproduced from *British Pharmacopoeia* (2007) vol. IV, appendix XIIF. A291 *Aerodynamic assessment of fine particles. Fine particle dose and particle size distribution.*

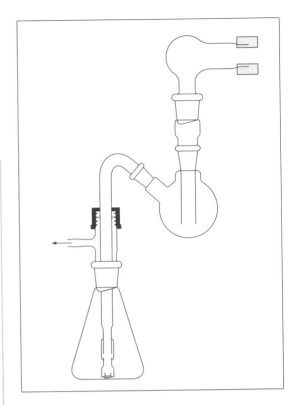

KeyPoints

- We have seen a selection of tests which can be conducted to measure the key parameters of a variety of formulations.
- These tests are not necessarily predictive of performance in vivo, but can be used in a comparative sense, testing one product against others or different batches of a product to ensure batch-to-batch consistency.
- Release tests can be applied to rectal and transdermal products by adapting the method used for oral products, altering the receptor phase to mimic the medium in which the formulation resides in vivo.
- Key parameters are different for different routes of delivery and different formulations: particle size is a key factor in inhalation products and in topical preparations where the drug is dispersed rather than dissolved in the vehicle.
- Adhesivity of oral dosage forms may be a factor in determining their efficacy (buccal delivery) or in causing adverse events (as in oesophageal injury); adhesion of transdermal patches to the skin is clearly important.
- The rheological properties of topical preparations and formulations for nasal delivery are important, and a key factor is the syringeability of injectables.

- There are several impaction surfaces at the back of the glass throat about 10 cm away from the activator (similar to human dimensions).
- The upper impinger (stage 1) has a cut-off at a particle size of ~6.4 μm.
- The last impact surface is in the lower impinger (stage 2) and is considered to be the respirable fraction.

Apparatus B

- Apparatus B is made of metal and can be engineered to finer tolerances than the glass apparatus A.
- Apparatus B is considered to be a superior apparatus for quality control testing and product release.

In vitro–in vivo correlations

- In vitro tests do not have to mimic the in vivo situation, but realistic parameters are important if useful data are to be achieved.
- The validity of the data from laboratory-based tests in predicting performance in the patient, however, depends on good in vitro–in vivo correlations.

Multiple choice questions

1. Indicate which of the following factors might lead you to consider applying dissolution testing to a pharmaceutical formulation:
   a. small particle size
   b. high drug solubility
   c. presence of polymorphs with poor dissolution characteristics
   d. evidence for a relationship between particle size and bioavailability

2. Indicate which of the following statements relating to in vitro testing methods are true:
   a. In natural convection methods there is no agitation.
   b. Forced convection methods cannot be used under sink conditions.
   c. Natural convection methods are representative of in vivo conditions.
   d. The British Pharmacopoiea rotating basket method is a forced convection method.

3. Indicate which of the following statements are true. Sink conditions:
   a. can be achieved by physical replacement of the solvent phase
   b. can be achieved by using a lipid phase into which the drug can partition
   c. allow the drug to achieve saturation levels in the dissolution medium
   d. should be used for drugs of very low aqueous solubility

4. Indicate which of the following statements relating to the adhesion of coated tablets to mucosal surfaces are true. The force of adhesion:
   a. is independent of film thickness
   b. depends on the rate of coat hydration
   c. depends on the hydrophobicity of the film coat
   d. is independent of the nature of the contacting surface

5. Indicate which of the following statements relating to the analysis of particle size distribution in aerosols using the cascade impactor are correct. This apparatus:
   a. is routinely used to monitor the air in clean rooms
   b. is a dynamic method of sizing
   c. separates particles by impingement on to baffles
   d. depends on both sedimentation and inertial forces

Answers to self-assessment

Chapter 1

1. b.
2. a and d.
3. c.
4. b and d.
5. a and c.
6. d and e.
7. a False.
   b True.
   c True.
   d False.
   e True.
8. a False.
   b False.
   c True.
   d False.
   e. True.

Chapter 2

1. b and d.
2. a, b and d.
3. a and d.
4. a, c, d and e.
5. b and d.
6. b.
7. c.
8. b.
9. c.
10. a and c.
11. d.
12. b and d.
13. a, c and e.

Chapter 3

1. a True.
   b False.
   c False.
   d True.
   e False.
2. a True.
   b True.
   c False.
   d False.
3. a False.
   b False.
   c False.
   d True.
   e False.
4. a False.
   b False.
   c True.
   d False.
   e True.
5. a True.
   b False.
   c False.
   d True.
   e True.
6. a True.
   b False.
   c False.
   d True.
   e False.
7. b and c.
8. d.
9. b.

Chapter 4

1. b.
2. a.
3. a False.
   b True.
   c False.
   d True.
   e True.
4. a True.
   b False.
   c False.
   d False.
   e True.
5. a, b and e.
6. a True.
   b False.
   c False.
   d True.
   e False.
7. a False.
   b False.
   c True.
   d False.
   e True.
8. a False.
   b True.
   c True.
   d True.
   e False.
9. b.
10. c.

Chapter 5

1. a and b.
2. c.
3. b.
4. b and d.
5. b, c and d.
6. a and c.
7. c.
8. a, c and d.
9. a and b.
10. c.

Chapter 6

1. a False.
   b False.
   c True.
2. a False.
   b False.
   c False.
   d True.
   e True.
3. a False.
   b False.
   c True.
   d True.
   e True.
4. a False.
   b True.
   c False.
   d True.
   e True.

5. a False.
   b True.
   c False.
   d True.
   e True.
6. a True.
   b False.
   c True.
   d False.
   e False.
7. a True.
   b False.
   c False.
   d True.
   e False.

Chapter 7

1. c and d.
2. b and d.
3. b.
4. c, d, f and g.
5. a, c and e.
6. a, d and e.
7. a, b, d and e.
8. a and b.
9. a and d.
10. d and e.

Chapter 8

1. a, b and e.
2. c and d.
3. b and d.
4. b and d.
5. a.
6. a and b.
7. a.
8. c.

Chapter 9

1. a and b.
2. a and d.
3. a, d and e.
4. a, b and d.
5. a and d.

Chapter 10

1. c and d.
2. a and d.
3. a, b and d.
4. b and c.
5. b, c and d.

Memory Diagrams

The diagrams can be further expanded in revision to give explanations of and connections to the phenomena displayed. Not all connections are shown. For example, in the first diagram what property of the drug does temperature influence? Particle size, crystal form or saturation solubility of drug?

Dissolution from a powder/suspension of drug or drug granules

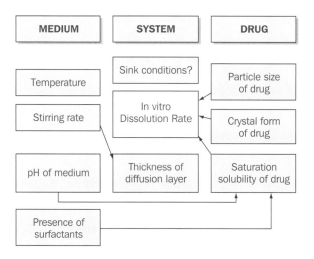

Dissolution from a disintegrating tablet (cf a matrix tablet)

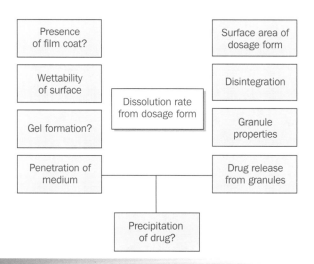

Release of drug from polymer matrices

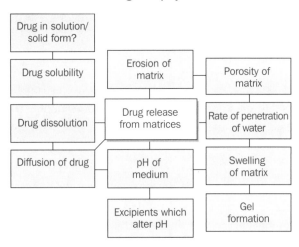

Properties of polymers

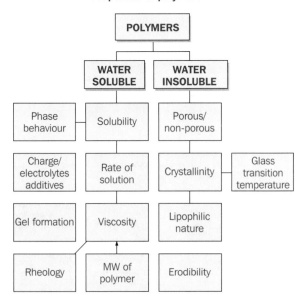

Chemical and physical factors affecting the stability of proteins

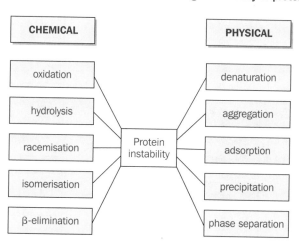

Stability of a hydrophobic colloidal system in an aqueous medium

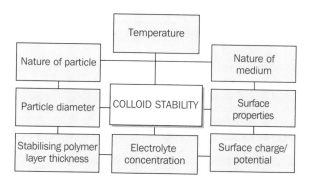

Factors affecting drug stability

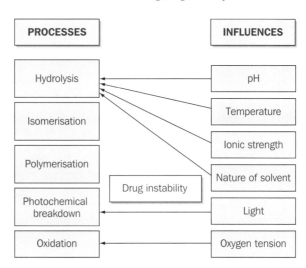

Factors affecting solubilisation in surfactant micelles

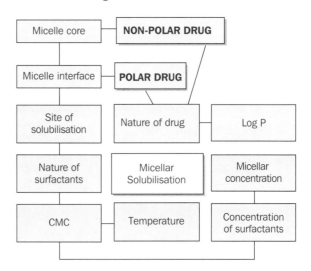

Classification and properties of surfactants

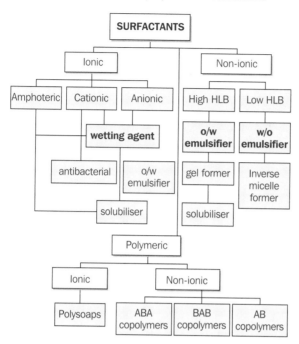

Properties of surfactants

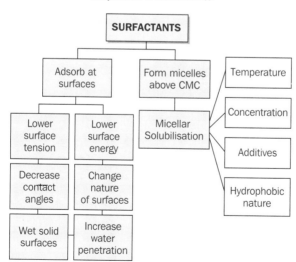

Features and means of *in vitro* testing of pressurised inhalers

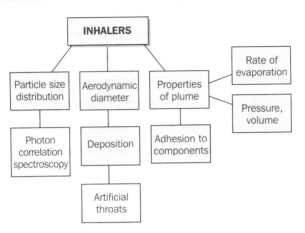

Index

Page numbers in *italic* refer to figures or tables.